Science Uncoiled

Michael T Deans

Published by

MELROSE BOOKS

An Imprint of Melrose Press Limited
St Thomas Place, Ely
Cambridgeshire
CB7 4GG, UK
www.melrosebooks.co.uk

FIRST EDITION

Cover designed by Melrose Books

978-1-910792-99-5
978-1-910792-13-1 epub
978-1-910792-93-3 mobi

Printed and bound in Great Britain by:
Lightning Source UK Ltd, Chapter House, Pitfield
Kiln Farm, Milton Keynes, MK11 3LW

I wish you goodness, truth, beauty, peace, love, progress, stability, justice and unity.

Michael T Deans

Contents

Foreword

Around 3.9 billion years ago, RNA and DNA molecules arose, during a primordial ice age, strange ice crystals formed in pools of liquid nitrogen on the Earth's poles. They released infrared laser light beams on warming and cooling; reflected and polarized by ice in clouds and on land and sea, some light reached the warm equatorial waters Darwin described in *On the Origin of Species*.

Nucleic acid molecules were concentrated by evaporation and absorbed it, combining with phosphates to create a noodle soup of DNA. Some formed pumps, 'transport DNAs', The DNA triplets on transport DNAs pair with those on 'differentiation DNAs', encoding tissue diets in the same way RNA triplets on transfer RNAs pair with those on messenger RNAs encoding protein amino acid sequences. They filled primitive cells with life's constituents and still do so.

Life is organised like libraries, supermarkets and universities, the chips in our brains prefer a particular arrangement. By cooperating, communicating and sharing our abilities, ideas and aspirations, mankind can offer the world a peaceful, loving and exciting future.

Part 1 considers how DNA stores information, patterns of hydrogen bonds connecting DNA to protein store memories, they switch between O ▯ H••••N and O••••H ▯ N on coiled abaci, those in any cell could hold the Bible, Koran and works of Shakespeare.

Part 2 discusses how our bodies use food, allowing us to know what's good, distinguish truth from falsity, appreciate the beauty around us, make peace, love everyone, progress life's diversity, ensure sustainability, administer justice and unify nations, philosophies and religions.

Part 3 describes how life started, that time's a figment of our imaginations and how my hopes and those of Newton, Darwin, Faraday, Heisenberg, Einstein, Pauling, Christ, Ghandi and Churchill could be realized.

Consequences

This beginning sets standards for change in any aspect of life, predicting the turmoil inappropriate developments create and prescribing solutions. Genetic engineers and drug designers need respect its alphabet and grammar. Expenditure on SETI, CERN and Tokomaks must be justified, specialist scientists need present acceptable public profiles. Entrepreneurs, politicians and theologians need respect minion limitations when resolving disputes arising from their projects, policies and beliefs.

The laws of physics need small adjustments to align them with minion logic. Particle physics, quantum mechanics and cosmology need review. Mathematicians shouldn't invoke zeroes and infinities. Engineers, architects and energy providers designing vehicles, housing and fusion reactors should study biological precedents.

Authorities regulating food standards should check trace element content, medical professionals broaden their understanding and user friendly computers emulate minions. School and college curricula ought to reflect my proposed disentangled science. Artists, fashion designers and composers might celebrate its simplicity.

Part 1: Only human

Preface

Stephen Hawking's *Brief History of Time* introduces a physicist's concept of time. Historians recognize cycles in social progress. The different mindsets prevailing in the Roman Empire's gladiatorial contests, the Victorian Empire's military conquests and current hopes for World Peace reflect two-thousand year cycles in human thought patterns. Studying them tells us something about how our minds work.

This Part combines established ideas with ancient traditions and some new ones. By examining ourselves, we can improve our lot and imagine a better future without regretting our failures. Greek and Roman cultures cultivated stability, justice and unity. Plato's goodness, truth, and beauty dominated Western culture from 0 $_{AD}$ to 2000 $_{AD}$. The major religions, including Judaism, Christianity, Islam and Buddhism, preach goodness and truth and sponsored the beauty of Renaissance art.

Peace, love, and progress are popular aspirations. I hope my proposals will foster them. This is a philosophy, challenging scientific dogma with a new explanation for the origin of life and how our minds work. We're only human.

1 Molecular intelligence

1.1 Genetic code

Part 3 describes how molecules of deoxyribonucleic acid, DNA arose. Crystals of a novel form of ice grew in pools of liquid nitrogen during a primordial ice age, releasing beams of infrared laser light of wavelength $\sim 4\mu$, $\mu = 10^{-6}$ metres when changing shape on warming and cooling. Repeated reflection by ice in clouds and on the Earth's surface shone polarized light on warm equatorial pools of water, where the electromagnetic energy added a phosphate group to deoxyadenosine diphosphate, dADP molecules, converting them to deoxyadenosine triphosphate, dATP (shown □ in Fig 1.1).

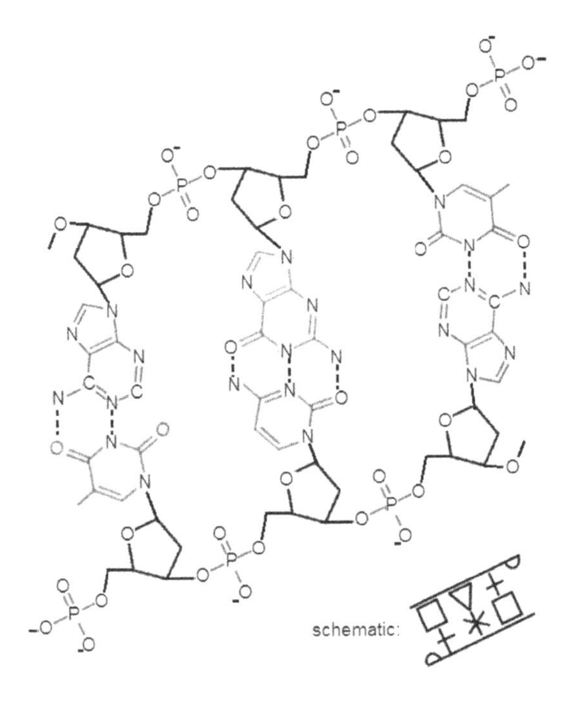

Fig 1.1 Flat DNA triplet

dATP combined randomly with similar molecules, dTTP, dGTP and dCTP (shown +, Δ and *), creating a noodle soup of DNA. Some DNA formed molecular pumps, transport DNAs, tDNAs, feeding primitive cells with small molecules, enabling their growth and reproduction, see Part 2.

I next consider three types of information storage on DNA: DNA triplet codes determine cells' diets, changing as they differentiate, forming different tissues. Triplets specify the amino acid sequences of proteins catalyzing metabolism throughout life. Arrangements of the hydrogen bonds, H-bonds binding DNA to protein store memories.

Transport DNAs tDNAs are three base-pairs wide, their three-base tags pairing with tags on the differentiation DNAs, dDNAs selecting them. Transfer RNAs, tRNAs, have three-base tags pairing with tags on the messenger RNAs, mRNAs specifying protein amino acid sequences. Proton ordered H-bond networks pointing in the same direction – all O–H••••N, not randomly O–H••••N or O••••H–N – store intellectual information, resonating, *ringing a bell* for memory recall.

History

In 1866, Gregor Mendel, an Austrian monk, described how peas he grew in his garden passed characteristics such as height and colour from one generation to the next; his findings were ignored. Twelve years later, Walther Flemming, a German scientist, proposed that the dark-staining chromosomes in cell nuclei were involved in cell replication; this idea was also ignored. These discoveries were the foundations of the science of inheritance, genetics.

X-ray diffraction

In 1953, James Watson and Francis Crick analyzed Rosalind Franklin's X-Ray diffraction images of DNA, obtained by shining X-rays into DNA extracts diffracting, bending the rays, creating patterns on photographic plates. They interpreted an X in the pattern to imply the

DNA molecules were helical. They determined the pitch of the helix was 34 Å, Å = Ångstrom, 10^{-10} metres, with base pair spacing 3.4 Å, proposing that adenine pairs with thymine, □ with + and cytosine with guanine, Δ with *, connected by H-bonds. Ribose sugar and phosphate connect these four component bases – adenine, cytosine, guanine, and thymine, □, +, Δ and * – in opposing strands, see Fig 1.1. They also proposed the genetic code, a 64-word language of base-pair triplets, which ribosomes, tangles of RNA, join forming amino acid protein sequences, Fig 1.2.

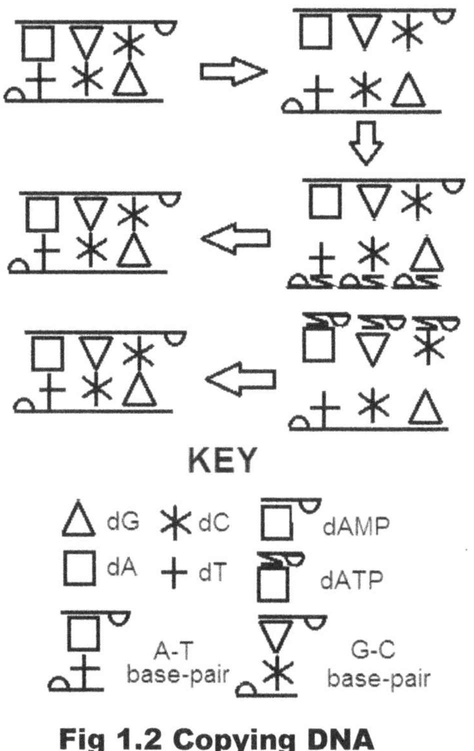

Fig 1.2 Copying DNA

1.2 Double helix?

This book is called *Science Uncoiled* after my proposal that DNA isn't coiled in living cells. If all the 2.7 B base pairs, 270 M turns of DNA in a cell were uncoiled, it would be two metres long. According to the consensus view of chromosome replication, copying involves breaking it into many fragments, taking about eight hours.

Sections unwind at 1,000 revolutions per second, split 10,000 base pairs per second and copy 20,000 base pairs, then rewind at 1,000 revolutions per second and the fragments magically reassemble correctly. The minion, the 'chip in the brain' enables copying with less likelihood of mistakes, mutations, arising. The double helix, Fig 1.3, uncoils, Fig 1.4:

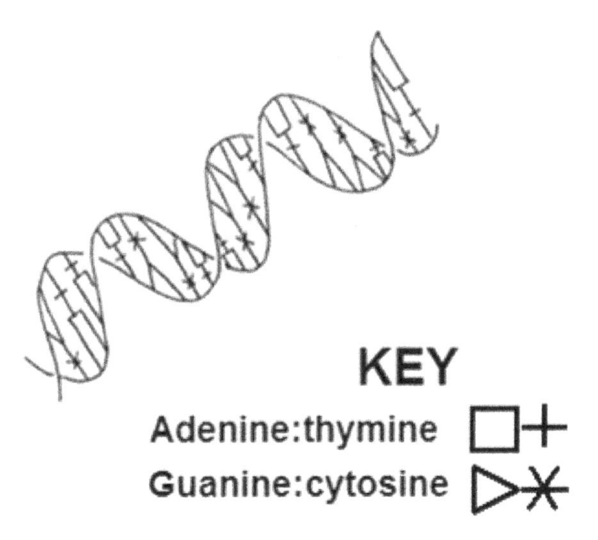

Fig 1.3 DNA double helix

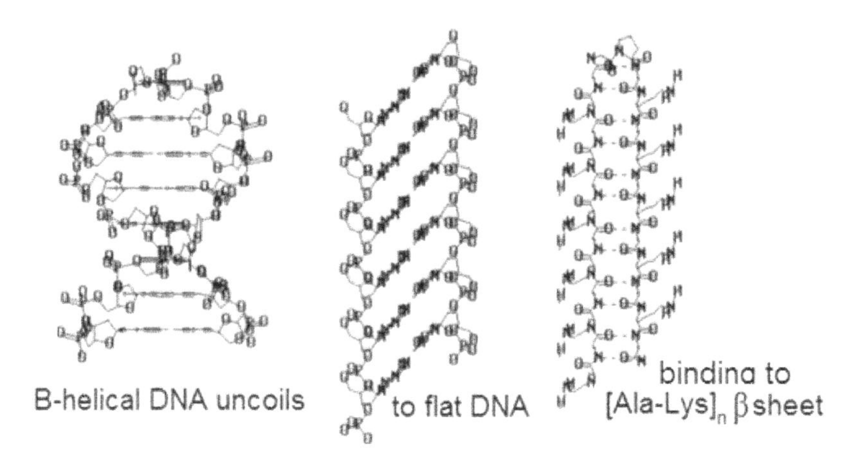

Fig 1.4 Uncoiled DNA fits β sheet

1.3 The role of protein

Histones

Histone proteins hold DNA flat, neutralizing it's acidity. They're usually composed of four neutral amino acids, ααs: alanine, leucine, isoleucine and valine, Ala, Leu, Ileu and Val alternating with two basic ααs: lysine, arginine, Lys, Arg. Proline, Pro, forms an asymmetric hairpin bend, making an anti-parallel β-pleated sheet, first proposed by Linus Pauling. Each histone makes a 17° angle with the next, 21 subunits form a coil, see Figs 1.5 and 1.6.

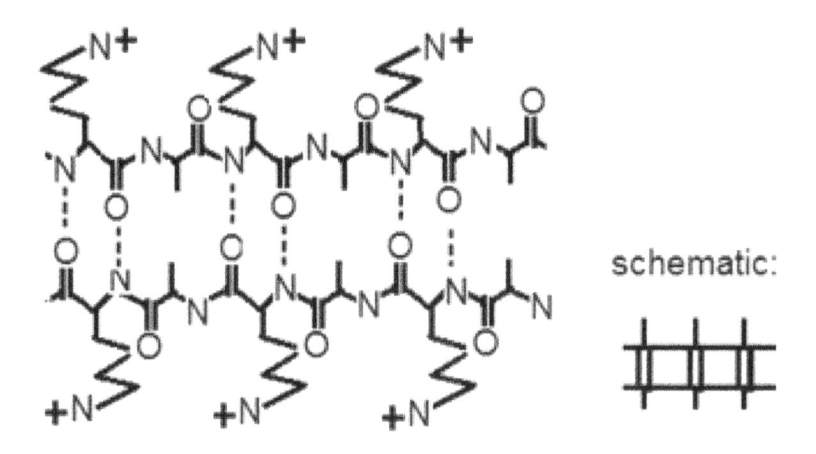

Fig 1.5 Protein β-sheet

Lys and Arg have positively charged ends, ω-amines binding histones to negatively charged phosphates, P_i on DNA. The base-pair overlap and spacing found in the double helix is retained. More H-bonds hold 1,701 base-pairs together, forming a 'minion' with nine coils, packing chromosomes better than the standard 'nucleosome core particles'. Cylindrical stacks ~2μ in circumference, μ = micron or 10^{-6} metres, super-coil forming chromosomes, Fig 1.7.

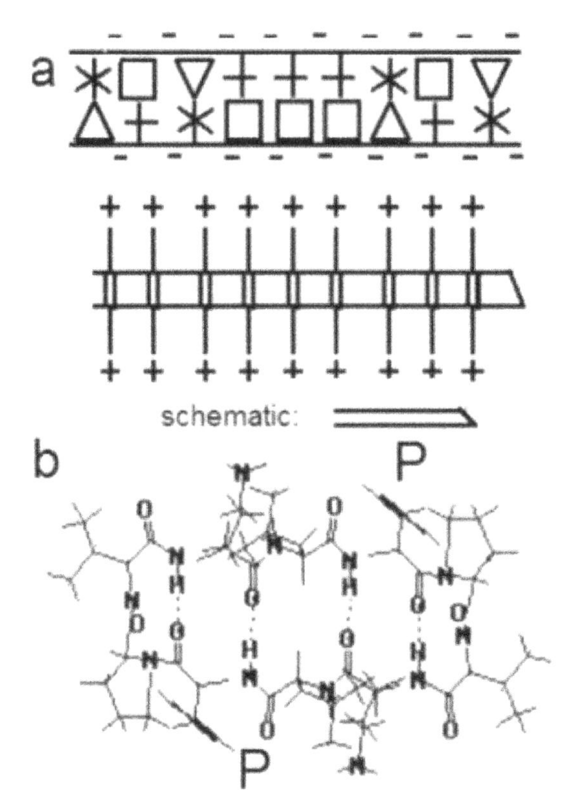

Bacterial protein Gramicidin S has an analogous structure, the phenlyalanine groups, P are equivalent to DNA bases.

Fig 1.6 Hairpin unit of protein binds DNA, Gramicidin S is analogous

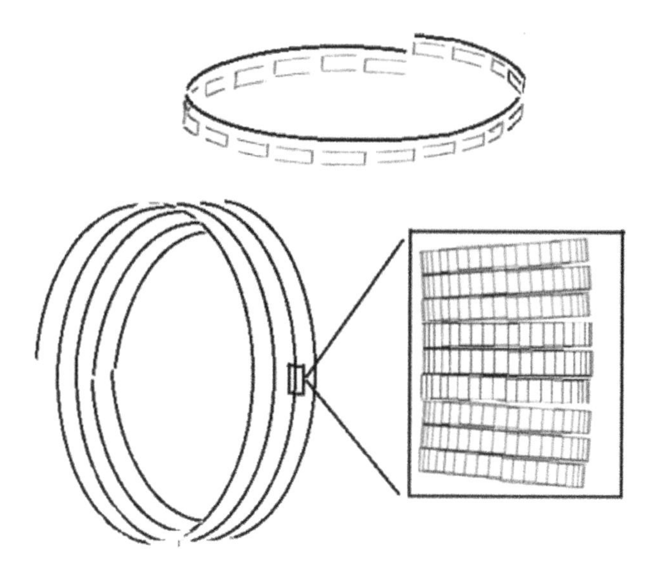

Fig 1.7 DNA packed in minions to form chromosome

The neutral amino acids: Ala, Leu, Ileu and Val fit specific DNA bases: Cytosine, Guanine, Adenine and Thymine respectively, mnemonic A LIVe CiGAreTte, Fig 1.8 shows how.

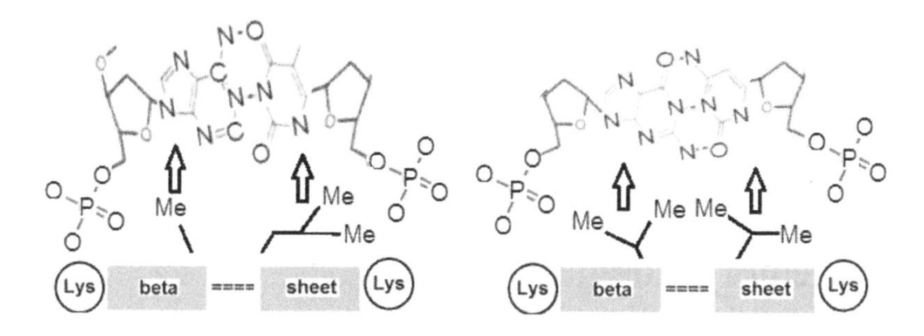

Fig 1.8 Matching amino acids with DNA base pairs

Histones are 3 DNA triplets long, the same length as recently studied 'siRNAs'. More histone proteins strap nine coils together forming a minion; the arrangement of these H-bonds accounts for memory storage. Fig 1.9 shows minions replicating 1701 base-pairs without any breaking and reconnection:

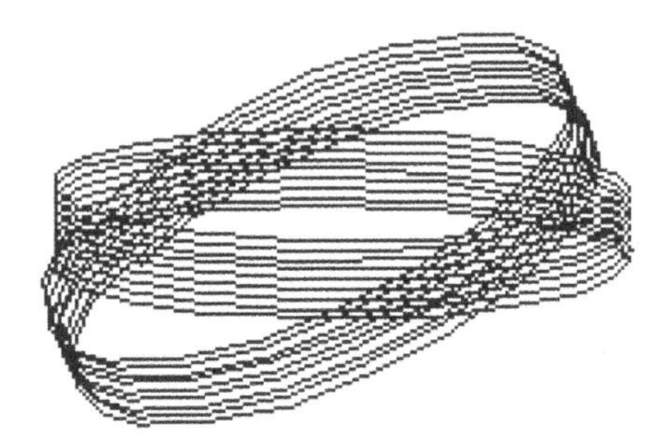

Fig 1.9 Minions replicate 1701 base pairs

Evolution

If errors, mutations, occur when copying sperm or egg cell DNA the progeny differ from their parents, beneficial changes signify evolution.

Proteins' second role was binding similar cells together in tissues. dDNAs are disabled during tissue differentiation, their associated tDNAs switch from substrate transport to protein synthesis producing matching proteins to 'hook' them together. Their third role was building structures like centrioles, the bundles of nine cylinders feeding energy along the spindles, triple α-helical coils of proteins, driving chromosome separation at cell division. Their role as catalysts may have emerged last. Some viruses have chromosomes made of RNA instead of DNA.

1.4 Memory storage

Because protein synthesis inhibitors weaken memory, many scientists believe memories are stored as different proteins, they don't offer a working mechanism. This model is inadequate: to encode proteins

50-300 residues long requires 50-300 DNA triplets. Only about 100 million thoughts could be stored.

Neural networks

Neural networks are the most popular account of memory: neuroscientists believe the route nerve signals follow is reinforced each time it's used. They also suggest groups of nerve cells exchange signals to process different types of information.

Subsequent encounters with the same input signals revive established pathways. Exercise reinforces memories, *e.g.* practicing a piano piece. Cell groups and connecting paths change when new experiences 'rewire' the brain, causing old memories to fade. More memories can be kept on neural networks than proteins, up to 100 trillion, 10^{11} pathways are available but recalling childhood memories in old age using this model seems improbable.

Holographs

Neither protein storage nor neural networks could store as many memories as minions. Both fail to explain the limited memory loss arising from destroying portions of the brain. The minion model of memory storage is 'holographic', every part of a holographic plate stores the whole image; every splinter of a broken plate contains a copy of the picture. Memory survives partial destruction without loss because copies of each minion memory store are dispersed throughout the brain. Both standard models lose memories irretrievably after brain damage, as surely as removing a page from a photo album.

Minions

'Minion', connotes memory and subservience, minions store architectural and intellectual information in parallel. Each coil has 21 subunits, 63 triplets or 189 base pairs. The H-bonds connecting base pairs oscillate, **a.** A 4μ infrared quantum activates an amine-phosphate

H-bond on the lower deck, **b** it's passed between decks, **c** and **d** then to an amine-phosphate bond on the upper deck. It's passed to the following amine-phosphate bond, **e** before departing,**f**.

Minions store standing waves on both their inner and outer coil surfaces distinguishing positive and negative or extrovert and introvert memories. Each of a minion's 18 coils stores one of 63 values, allowing $63^{18} \approx 2.44 \times 10^{32}$ possible eighteen character words. Combinations form memory sentences, their meanings independent of location, see header cartoon and Fig 1.10:

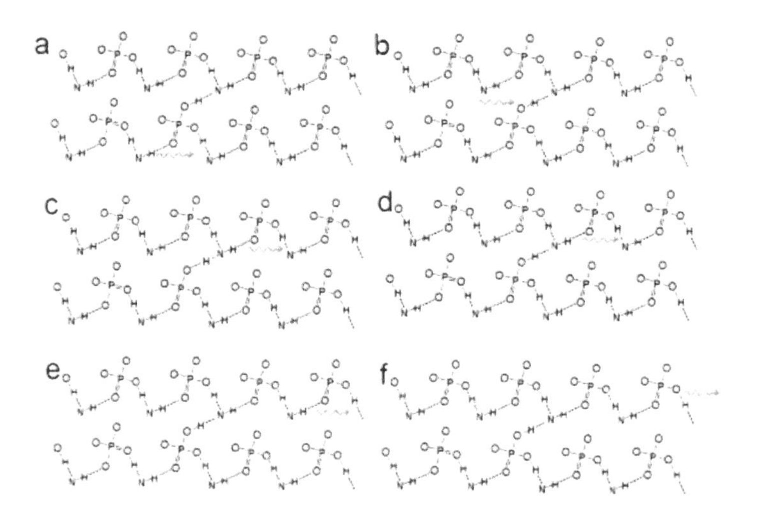

Fig 1.10 Domino hydrogen bond switching

1.5 Counting

The beads on an abacus count 1 to 9 on the first row, 1 to 9,000 on the fourth, enabling multi-digital calculations. A 10-row abacus can count up to 10^{10}. Minion coils count on base 63, the first coil's triplets represent numbers 1 to 63, those on the second 63 to 63*63 = 3969, both its surfaces count up to 63^{18}. Dekatrons™ use similar logic to count on base 10, Fig 1.11, Table 1.1 lists powers of 63.

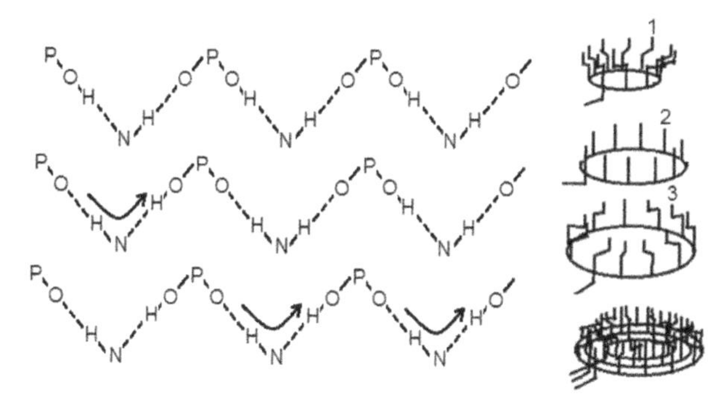

Fig 1.11 Minion and Dekatron™ counting compared

$$63 = 63^{1}$$
$$3969 = 63^{2}$$
$$250047 = 63^{3}$$
$$15752961 = 63^{4}$$
$$992436543 = 63^{5}$$
$$62523502209 = 63^{6}$$
$$3938980639167 = 63^{7}$$
$$248155780267521 = 63^{8}$$
$$15633814156853823 = 63^{9}$$
$$984930291881790849 = 63^{10}$$
$$62050608388552823487 = 63^{11}$$
$$3909188328478827879681 = 63^{12}$$
$$246278864694166156419903 = 63^{13}$$
$$15515568475732467854453889 = 63^{14}$$
$$977480813971145474830595007 = 63^{15}$$
$$61581291280182164914327485441 = 63^{16}$$
$$3879621350651476389602631582783 = 63^{17}$$
$$244416145091043012544965789715329 = 63^{18}$$

or approximately, $63^{18} = 2.4 \times 10^{32}$

Table 1.1 Powers of 63

Standing waves

Standing waves keep their positions relative to the medium, *e.g.* if a couple hold a skipping rope and the husband shakes his end, a wave travels toward his wife; when she reciprocates, the wave oscillates or stands. To understand minion coil counting, imagine standing waves 64 base pair triplets long, slightly more than a coil.

The simplest standing waves have fewest stationary nodes, they fall between triplets. Fig 1.12 shows waves with four, eight 16 and 32 nodes. Fig 1.13 illustrates counting, involving base-pair triplets labelled COD and histones labelled 1 to 9.

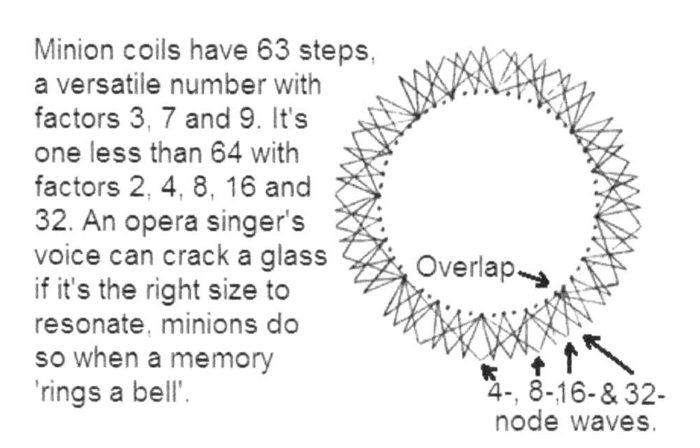

Fig 1.12 Resonance

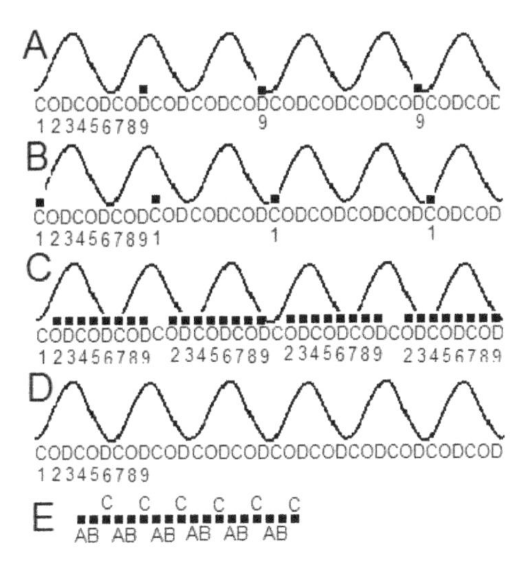

Fig 1.13 Dekatron™-like wave propagation

By this analogy, standing waves two triplets long count as zero; a nerve stimulus excites the histone marked *, shifting the wave right, leaving an unstable node between C and O. A second stimulus shifts it further to between O and D, a third restores it to a stable position with its nodes again on triplet boundaries. The wave continues to count one until some disturbance triggers counting two. A coil's circumference is $\sim 0.139\mu$, so light takes about 4.6×10^{-16} seconds to pass round and back and establish a standing wave. Each of the three steps per count takes $\tau = 1.39 \times 10^{-15}$ seconds, τ = Greek letter tau, τ limits conception, replacing the physicists' Planck's constant, h.

1.6 Boundaries of mind

Our minds can comprehend neither the infinite nor the infinitesimal, understanding nothing outside the minion's frame of reference. Infinity lies beyond the largest conceivable number, nothing smaller than an electron is imaginable.

Time, distance and mass

$\tau = 1.39 * 10^{-15}$ seconds, the smallest conceivable time period, features in both particle physics and cosmology. Our psychological reaction time, $\tau * 63^9 \approx 21$ seconds is the largest the minion's inner surface can contemplate. The outer can conceive the age of the universe, about $\tau * 63^{18} \approx 10.8$ billion years, Tables 1.2, 1.3 and 1.4 list the times, distances and masses corresponding to its 18 coils.

overflow		$2.9 * 10^{17} = 9{,}200$ M year
9th		$4.6 * 10^{15} = 150$ M year
8th	F	$7.3 * 10^{13} = 2.3$ M year
7th	A	$1.2 * 10^{12} = 37{,}000$ year
6th	C	$1.8 * 10^{10} = 590$ year
5th	E	$2.9 * 10^{8} = 9.4$ year
4th	a	$4.7 * 10^{6} = 54$ day
3rd		$7.4 * 10^{4} = 21$ hour
2nd		$1.2 * 10^{3} = 20$ min
1st		$1.9 * 10^{1} = 19$ sec
1st		$3.0 * 10^{-1} = 0.3$ sec
2nd	F	$4.7 * 10^{-3} = 4.7$ m sec
3rd	A	$7.5 * 10^{-5} = 75$ μ sec
4th	C	$1.2 * 10^{-6} = 1.2$ μ sec
5th	E	$1.9 * 10^{-8} = 0.019$ μ sec
6th	b	$3.0 * 10^{-10} = 300$ μμ sec
7th		$4.8 * 10^{-12} = 4.8$ μμ sec
8th		$7.6 * 10^{-14} = 0.0076$ μμ sec
9th		$1.2 * 10^{-15} = 0.0012$ μμ sec

Table 1.2 Minion coil times

overflow	$8.8 * 10^{25} = 9{,}300$M light yr
1st F	$1.4 * 10^{24} = 15$M light yr
2nd A	$2.2 * 10^{22} = 230{,}000$ light yr
3rd C	$3.5 * 10^{20} = 3{,}700$ light year
4th E	$5.5 * 10^{18} = 59$ light yr
5th a	$8.8 * 10^{16} = 0.93$ light yr
6th overflow	$1.4 * 10^{15} = 9{,}400$ ast units
7th 9th	$2.2 * 10^{13} = 150$ ast units
8th 8th	$3.5 * 10^{11} = 2.3$ ast units
9th 7th F	$5.7 * 10^{9} = 5.7$M km
6th A	$8.9 * 10^{7} = 89{,}000$ km
5th C	$1.4 * 10^{6} = 1{,}400$ km
4th E	$2.3 * 10^{4} = 23$ km
3rd b	$3.6 * 10^{2} = 360$ meter
2nd	$5.7 * 10^{0} = 5.7$ meter
1st	$9.0 * 10^{-2} = 9$ cm
1st	$1.4 * 10^{-3} = 1.4$ mm
2nd	$2.3 * 10^{-5} = 22$ µm
3rd F	$3.6 * 10^{-7} = 0.36$ µM
4th A	$5.7 * 10^{-9} = 57$ Å
5th C	$9.0 * 10^{-11} = 0.9$ Å
6th E	$1.4 * 10^{-12} = 1.4$ µµM
7th a	$2.2 * 10^{-14} = 0.022$ µµM
8th	$3.6 * 10^{-16} = 0.00036$ µµM
9th	$5.8 * 10^{-18} = 5.8$ µµµM

Light year = distance light travels in a year, Å = 10^{-10} m,
ast (Astronomical unit) = Earth-Sun distance, µ = 10^{-6}

Table 1.3 Minion coil distances

overflow		$2.0 * 10^{33}$ = mass of Sun
9th		$5.0 * 10^{29}$ = 84 Earth masses
8th	F	$1.3 * 10^{26}$ = 1.8 Moon masses
7th	A	$3.1 * 10^{22}$ = 31 Pton
6th	C	$8.0 * 10^{18}$ = 8 Tton
5th	E	$2.0 * 10^{15}$ = 2 Gton
4th	a	$5.0 * 10^{11}$ = 500 kton
3rd		$1.3 * 10^{8}$ = 130 ton
2nd		$3.2 * 10^{4}$ = 32 kg
1st		$8.1 * 10^{0}$ = 8.1 g
1st		$2.0 * 10^{-3}$ = 2 mg
2nd	F	$5.1 * 10^{-7}$ = 0.51 μg
3rd	A	$1.3 * 10^{-10}$ = 130 ng
4th	C	$3.3 * 10^{-14}$ = 0.033 pg
5th	E	$8.3 * 10^{-18}$ = 8.3 ag
6th	b	$2.1 * 10^{-21}$ = 2 base-pair masses
7th		$5.2 * 10^{-25}$ = $\frac{1}{3}$ proton mass
8th		$1.3 * 10^{-28}$ = $\frac{1}{7}$ electron mass
9th		$3.3 * 10^{-32}$ = $3 * 10^{-18}$ joule-equiv
		P, T, G, k, m, μ, n, p, a mean 10^{15}, 10^{12}, 10^{9}, 10^{3}, 10^{-3}, 10^{-6}, 10^{-9}, 10^{-12} and 10^{-18} respectively.
		Tiny masses are often referred to as energies.

Table 1.4 Minion coil masses

1.7 Balance

Imagination

The minion's structure limits our knowledge of reality; we can use ideas from education, environment and experience to imagine possibilities before doing anything. Diplomats drafting peace treaties might make better balanced decisions if they used minion logic to understand how their minds worked.

Belief

Depending on their personalities, education and environment, observers see things differently. People agreeing what they see, believe it, *e.g.* the Sun shines, rises in the east and sets in the west. Nicolaus Copernicus was outlawed for questioning the belief that the earth lay at the centre of the universe with everything revolving around it. Thinkers had long debated whether the Sun orbits Earth or *vice versa*. Mathematicians and astronomers find it simpler to regard the Earth as orbiting the Sun.

Thinking and acting

Logical thought and questioning observations are better at improving understanding of the world's working than random correlations. If we see someone whenever it rains, we shouldn't regard them as rainmakers. The world can be improved by changing our beliefs.

Ego

Balancing time spent believing, thinking and acting leads to the best decisions. Egos differ from clocks when recording time. Though your first kiss was a brief encounter, it's remembered for the rest of your life. Believing only common knowledge, inventing and communicating ways to ease others' lives makes for a happy ego.

1.8 The thought spectrum

Assigning meanings to numbers

Minions use the same way to start both linear and logarithmic counting. Fig 1.14 shows a stack of minions with the thought spectrum alternately green and red, inside and outside the cylinder, labelled with the twelve signs of the astrological zodiac representing different ways of thinking.

Part 3 introduces the physics of H-bond switching and relativity between perception and conception. The zodiac number, Z = 2190, three more than the thought spectrum's 2187 steps, correctly predicts the number of days in a year (365.24219879):

$$\frac{2190/6 + 1/4 - 18/2190 + 11/12 - 3/2190}{2190} = 365.242198766$$

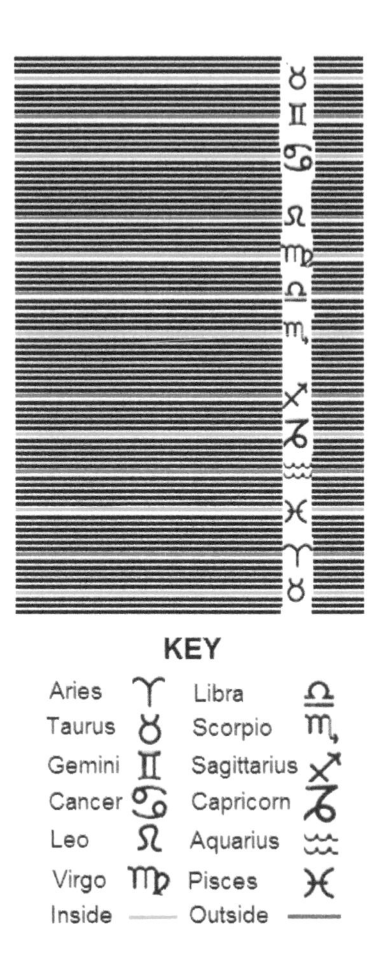

Fig 1.14 Astrological zodiac

Day and year length can be derived from the time unit, τ, Table 1.5. About every two thousand years, the equinoxes change, Table 1.6 shows the configurations for the past Age and the New Age we're entering.

Day = $2 * 63^9 * (2190/6)^2 * 1.32 * 10^{-15}$ seconds ≈ 24 hours
Year = $2 * 63^9 * (2190/6)^3 * 1.32 * 10^{-15}$ seconds ≈ 365 days

Table 1.5 Day and year

Old age	+1 −7 +4 −1 +2 −8 +5 −2 +3 −9 +6 −3
New age	+1 −7 +4 −5 +2 −8 +5 −6 +3 −9 +6 −4

Table 1.6 Age structures

During the New Age, coils 4, 5 and 6 replace 1, 2 and 3. Peace, love and progress substitute for the Platonic virtues, goodness, truth and beauty. The minion model confirms astrologers' belief that the rising and Sun signs at birth, changing twelve times each day and year, influence an individual's thought processes throughout life, presuming that the first breath etches a datum in every minion.

1.9 Nine ways of thinking

Comparing coils

We should balance belief and thought when making decisions. A café's lunch menu features salad, meat and two vegetables, sandwiches or risotto, some prefer the freshness of salad, others the taste of meat, the convenience of sandwiches or high calorie risotto. Everyone would choose salad if it was better in every respect, but we use our nine minion coils when deciding anything, whether what to eat for lunch or to declare war, each thinking differently:

1 Goodness – children learn to be good, what's good or bad is a subset of what's right or wrong.

2 Truth – our morals are based on what we believe to be true, truth is distinct from goodness, what's true may be bad, what's good false.

3 Beauty – easily diverts attention, we use it to evaluate works of art and marriage partners.

4 Peace – all religions advocate peace, peace of mind prepares us for making decisions. Educators should provide a breadth of knowledge to nurture curiosity and ingenuity.

5 Love – makes the world go round, we're essentially social animals needing companionship and affection.

6 Progress – scholars thinking outside the box, inventors exploring new ground and scientists conducting green fields research fuel progress. Research and development are the gilt on the New Age gingerbread.

7 Stability – it's difficult to accomplish anything if food and shelter are uncertain; a stable home, steady income and supportive social group are everyone's right.

8 Justice – crime, corruption and prejudice make life difficult, justice creates a level playing field by rehabilitating offenders, prosecuting fraudsters and assuring equal opportunity.

9 Unity – life is easier when governments, trades unions and employers provide a consistent, united front.

These nine ways of thinking work best in parallel, correcting personal bias by reference to others' ways of thinking. The ways are independent, a mathematician would call them orthogonal, weighing equally in decision making, Table 1.7.

0 maths, ego 0						
psychology	2 seconds	-1	goodness	+1	psychiatry	3 minutes
technology	$^{1}_{30}$ seconds	-2	truth	+2	management	3 hours
biology	$^{1}_{2000}$ seconds	-3	beauty	+3	sociology	1 week
genetics	5 μμgrams	-4	peace	+4	politics	1 year
electronics	1 MHz	-5	love	+5	history	75 years
biochemistry	molecular masses	-6	progress	+6	archaeology	5,000 yrs
chemistry	atomic mass	-7	stability	+7	palaeontology	300,000 yr
physics	electron mass	-8	justice	+8	astronomy	planet mass t□mass
particles phys.	high energies	-9	unity	+9	cosmology	Sun's mass

Table 1.7 Mental associations with minion coils

2 Through the mind's eye

If a thought isn't immediately useful, it may be in the future. Our attitudes are determined by the way nature, nurture and nativity preset our minions. This chapter focuses on understanding them. In Chapter 3, I suggest ways to achieve agreement through cooperation, collaboration and communication despite our different mindsets.

2.1 Numbers of things

CONSTANT	SYMBOL	VALUE
Planck time	t_P	$5.391 * 10^{-44}$
Planck length	l_P	$1.616 * 10^{-35}$
Planck's constant	h	$6.626 * 10^{-34}$
Nuclear magneton	μ_N	$5.051 * 10^{-27}$
Electron mass	m_e	$9.109 * 10^{-27}$
Proton mass	p_m	$1.673 * 10^{-26}$
Bohr magneton	μ_B	$9.274 * 10^{-24}$
Boltzmann constant	k	$1.381 * 10^{-23}$
Bohr radius	α_0	$5.292 * 10^{-21}$
Elementary charge	ϵ	$1.602 * 10^{-19}$
Magnetic flux quantum	Φ_0	$2.068 * 10^{-18}$
Hartree energy	E_h	$4.346 * 10^{-18}$
Electric constant	ϵ_0	$8.854 * 10^{-12}$
Newton's gravity const	G	$6.674 * 10^{-11}$
Planck mass	m_P	$2.176 * 10^{-8}$
Magnetic constant	μ_0	$1.257 * 10^{-6}$
Fine structure constant	A	$7.297 * 10^{-3}$
Impedance of vacuum	Z_0	$3.767 * 10^{2}$
Von Klitzing constant	R_K	$2.581 * 10^{4}$
Rydberg constant	R_∞	$1.097 * 10^{7}$
Speed of light	c	$2.998 * 10^{8}$
Josephson constant	K_2	$4.836 * 10^{14}$
Avogadro number	A_N	$6.022 * 10^{23}$

Table 2.1 Fundamental constants

ln^{-1}	N	ln^{-1}	N	1 digits									1 digits								
8.09	123000000	7.11.12	13180000	1	2	3	4	5	6	7	8	9	1	2	3	4	5	6	7	8	9
2.64464	437	5.08	120200	1	2	3	4	5	6	7	8	9	1	2	3	4	5	6	7	8	9
2.31	204	6.01	3990000	1	2	3	4	5	6	7	8	9	1	2	3	4	5	6	7	8	
9.84	6918000000	6.86	3990000	1	2	3	4	5	6	7	8	9	1	2	3	4	5	6	7		
7.79	6166000	7.78	6026000	1	2		4	5		7	8	9	1	2	3	4	5	6			
9.12	9318000000	0.93	9	1	2		4	5		7	8	9	1	2	3	4					
4.58	3802	4.22	12600	1	2		4	5		7	8	9	1	2	3	4					
3.92	8318	7.64	4365000	1	2		4			7			1	2	3	4					
2.45	282	7.13	13490000	1	2		4			7			1	2	3	4					
2.07	118	3.59	3890		2					7			1	2	3	4					
4.47	49510	2.31	204							7			1	2	3						
0.93	9	3.86	7244							7			1	2							
8.04	109600000	7.76	5754000							7			1	2							
0.34	2	0.41	3										1	2							
4.82	66070	7.51	3236000										1	2							
7.17	14790000	5.73	537										1								
8.58	380200000	5.55	354800										1								
1.56	36	7.41	25700000										1								
8.95	891300000	4.53	33880																		
6.24	4207000	9.38	2399000000																		
1.28	19	7.19	15490000																		
8.68	48760000	5.85	7079																		
7.36	22910000	9.88	7586000000																		
2.63	427	4.38	43990																		
8.77	58800000	1.56	36																		
9.42	2630000000	9.82	6607000000																		
9.82	6607000000	1.28	19																		
8.61	40740000	6.47	2951000																		
0.20	2	1.52	33																		
1.95	89	2.67	470																		
8.75	562300000	2.69	490																		
0.12	1	5.25	177800																		
7.39	24550000	0.14	1																		

Table 2.2 Antilogarithms and first digits of random numbers

Relative numbers

The rules of arithmetic learnt in primary school don't apply to minions. If your favourite brand of chocolate is reduced from 10 to 8p or a magazine featuring a defence spending reduction from £1,000M to £800M is reduced from £1 to 80p, in the mind's eyes they're all 20% reductions; minion comparisons are relative.

Arithmetic balance

All daily activities can be allotted to minion coils. To assess your personality bias, compare the times you spend on each. If they're unequal, consider changing priorities. Numbers rating significantly less than others indicate aspects of life you've ignored. Greater happiness may be gained by balancing attention amongst coils.

Randomness

The fundamental constants of science in Table 2.1 aren't as random as they first seem. Although they range from 10^{-44} to 10^{23}, their first digits are more often 1 or 2 than 8 or 9. Their logarithms are random, not their absolute values. The random collection of numbers taken from the Telephone Directory in Table 2.2 with the first digits of their 'natural', base-e anti-logarithms, demonstrates the same exponential distribution. Since minion logic thinks logarithmically, most number-sets follow this pattern. Coil number takes precedence over coil position when minions compare numbers.

2.2 If I were you

Free will

If I thought as you do and did as you did, I wouldn't be using my free will. If you decided to have coffee and cake, so would I. You can learn about yourself by comparing your choices with those of others.

Perception

If after visiting a botanical garden with five friends, the gardener asks you to describe the garden in one word, you might answer *beautiful*, the others *abundant, tidy, aromatic, colourful* and *exciting*. The six answers reflect personality, education and experience. The friend replying *colourful* may not have noticed whether the garden was *aromatic*. You may not have registered the flowers' *abundance*. You can become more perceptive through discussion with friends.

Outlook

Our world view is influenced by experience, opportunity, and hope. Experience provides a frame of reference, customs and laws to respect, morals to follow and mishaps to learn from. Opportunity enables visualizing a better future. Hope encourages reaching for opportunity, without hope we're stuck with our habits. The key to a happy future is optimism.

2.3 Colours

The colours minions recognize can be established by comparing those used by advertising agencies, film makers, recommended by interior decorators and cited in folklore. Colours such as those used in nurseries keep us calm; others may enrage, excite or overpower us. Colours evoke emotions; those we choose for decorations reflect our personalities. Table 2.3 summarises colour associations with common phrases, Table 2.4 lists those related to minion coils.

being green	naive
golden rule	truthful
colourful	many sided
yellow with age	tranquil
warm brown	affectionate
royal purple	progressive
blue blooded	privileged
silvery moon	dreaming
red	socialist

Table 2.3 Colour associations

Use psychological colour associations as a guide when decision-making or negotiating. Repaint your room in appropriate colours. Show your true colours.

green	goodness	gold	truth	pied	beauty
yellow	peace	bronze	love	violet	progress
blue	stability	silver	justice	red	unity

Table 2.4 Colour associations

2.4 Relating

Communication

Having seen how the mind interprets numbers, colours and truths and sees, hears and believes what it likes, we consider how mental bias distorts meanings and relationships. The *three Rs*, Reading, wRiting, and aRithmetic and rules of grammar learnt as children provide a common language enabling effective communication, correcting errors and forming successful relationships.

Confusion

At first sight, the nine coils appear similar; to a lawyer, goodness, truth and justice may seem indistinguishable. We confuse truth with goodness when telling white lies to save a loved one's feelings and confuse unity with stability when seeking political compromise. To avoid confusion about ideas, the words for the nine coils need to mean the same to everyone.

Compromise

Successful relationships rest on compromise, shared understanding of the nine coils in the light of experience. It's difficult to compromise our moral values when building relationships, but small changes to our attitudes strengthen bonding. Insisting on my idea of goodness could disrupt our friendship, allowing for your point of view makes for a better relationship.

2.5 Occult

The minion model explains ancient traditions and beliefs.

Natal astrology

Section 1.8 mapped the connection between minion structure and the thought spectrum. The astrological Sun signs change monthly and every two hours. Perhaps the shock of an infant's first breath of air etches a mindset into all minions, determining choices throughout life.

Numerology

Numbers have always had meanings; numerologists relate number patterns and synchronicities in our lives to our personalities. Table 2.5 shows the meanings associated with the nine coils.

1 goodness	2 truth	3 beauty
4 peace	5 love	6 progress
7 stability	8 justice	9 unity

Table 2.5 Coil meanings

The success of numerology arises from minion structure. Our memory may be pictured as a library with nine categories, the books shelved according to their page count. To assign the books to categories, the librarian divides the page count by 9 and uses the remainder. 1 page leaflets go in the first category, 2 page leaflets in the second, etc; counts divisible by 9 belong in the ninth category. *E.g.* 245/9 = 27 remainder 2 belongs in the second category.

Fadic addition, adding digits together *e.g.*:

$$2+4+5 = 11,\ 1+1 = 2$$

easily finds the remainder for large numbers. To file a merged book, *e.g.* one with 23 pages from the fifth category combined with one with 112 pages from the fourth containing 23+112 = 135 pages is assigned to the ninth category. Another from the first category with 19 pages merged with 36 pages from the ninth totalling 19+36 = 55, 5+5 = 10, 1+0 = 1 belongs in the first.

Base-ten arithmetic originated with the minion, not our ten fingers. Using nine categories is the natural way to count, matching brain storage. Computers think in base two, binary, humans exceed digital machines in imagination, reasoning, emotions and other abilities.

Alphabet

Numerical values may be assigned to the letters of the alphabet, in his book, *Numera,* Bernard Spencer le Gette's proposed:

A	B	C	D	E	F	G	H	I	J	K	L	M	N	O	P	Q	R	S	T	U	V	W	X	Y	Z
1	2	2	3	4	5	8	5	1	1	2	3	4	5	7	8	1	2	3	4	6	6	6	5	1	7

MICHAEL translates to 4+1+3+5+1+5+3 = 22, 2+2 = 4, it belongs in the fourth category, corresponding to peace. Numerologists believe your given name matches your mental bias and reflects your personality. If I preferred to be called MIKE, my number would be 4+1+2+5 = 12, 1+2 = 3 in the third category, implying a preference for beauty, not peace. False names can be misleading.

2.6 Prophecy

Records

Many people have recorded what they've seen, where they've been, their hopes, histories, plans and predictions. A prophecy predicts future events. Weather forecasters base their predictions on collected information and architects' blueprints determine a building's appearance, they're often accurate. Realizations of the Biblical prophecy of Babylon's destruction or Nostradamus' visions of the French Revolution and rise of Hitler are bewildering.

Interpretation

When Biblical predictions or those of Nostradamus come true, it's disconcerting, but prophecies shouldn't be overrated. Without a frame of reference, they have many interpretations; the prophet has observed human nature and describes oft repeated scenarios. *E.g.* one of Nostradamus' quatrains was claimed to describe WWII fighter pilots with oxygen masks, yet powered flight was unknown when it was written. Prophets use metaphors; interpreting their predictions requires historical knowledge.

Plans

An architect's blueprint is a plan designed to satisfy his client. Political leaders realizing a prophecy may cause offence. Israel's rebirth isn't justified by Daniel's prediction. When Oedipus realized a prophecy by killing his father, it made him an outcast. Plans need democratic debate before they're implemented.

Statistics

Realizations of prophecies are merely examples of history repeating itself. Statistical analysis and predictions based on probability are limited, using two-way, not nine-way thinking. Statistics are a useful basis for planning, but don't justify plans.

Problem solving

Prophets' predictions reflecting all nine dimensions are useful for resolving government problems. Problems are best solved by using maps with nine orthogonal axes labelled -9 to +9, *e.g.* diplomats resolving difficulties arising from a peace treaty need take many matters into consideration. Historical precedents at the same location on the nine-dimensional map are useful guides.

2.7 Ego

The zero coil

Some critics view my nine ways as shades of grey, preferring to see everything in black and white. Their egos override all nine coils with an overarching 0 coil. Everyone is influenced by their egos, but needs all nine ways to reach a satisfactory compromise with others' viewpoints.

I could allow for my ego by seeing myself from your viewpoint, it's important to understand your ego and compensate to ensure decisions are rational, not egocentric. Our egos may be distorted by mental illness; when we lose sight of ourselves through illness or grief, psychiatrists and psychologists try to restore normality.

The Board

Imagine a boardroom with nine counsellors debating an unhappy circumstance, agreeing it was unfortunate. Their opinions will differ, either seeing a hopeless outcome, proposing solutions or offering comfort and advice to survivors. The imaginary board can solve problems if the only dissenting opinion is your own.

Egos seek to be godlike or happy, taking board members' advice facilitates choosing a happy outcome to satisfy your ego. As you get to know your internal or external counsellors, you'll gain increased confidence in their advice and your decisions. Helping others' egos to be happy can also satisfy yours.

2.8 God

Everything

My ideas are easily dismissed as a Theory of Everything, implying they're unlikely to be true. It's convenient to propose that our visions and revelations reflect a higher power or god. The idea of god need embrace everything, nothing can be ungodly, his seal of approval makes life happier. With some exceptions, such as weapons of war, innovations improve life.

My God

Everyone's idea of god is determined by experience, upbringing and personality. As an atheist, my idea of god as a human construct doesn't diminish my need to respect yours. Prayer, meditation and communal worship serve to balance our lives.

Utopia

Sharing happiness with others goes some way towards achieving a vision of utopia. Chapter 3 describes a future which I believe would make everyone happy and satisfy everyone's idea of god. Before utopia can be established, everyone need change and discover some happiness to share with others. Don't keep yours a secret.

2.9 Freedom

Freedom has been the battle cry throughout history. All the forms of government described in Plato's *The Republic* have been overthrown by citizens clamouring for freedom, claiming that it failed to satisfy their needs. The best offers each generation more freedom than its predecessor. Most modern governments balance rights and freedoms, affording greater freedom than Plato enjoyed. An ideal government would make everyone happy, free to be individually content. Eventually, society may need no government, but any breach of faith would resurrect it. I next examine ways to manage society and their implications.

3 New Age

The start of the Age of Aquarius, the New Age is disputed. Some believe it began in the 20th century, some say it's yet to come. Whenever it starts, this chapter is intended to prepare you for it, describing a new frame of mind and how a communal mindset can facilitate relationships and a happier society. It isn't an instruction manual but a catalogue to choose from. An Age is ending and ships bound for uncharted seas are being launched, don't miss the boat!

3.1 Reliance on goodness

Solitary animals need only care for themselves, eat, drink, mate and enjoy life. As a social species, mankind also needs to preserve society. Ants and honeybees make sacrifices for the common good, respecting the colony's welfare above their own. Individuals may be ignorant and prejudiced or have conflicting moral codes, challenging anyone disagreeing with them. Acting in our own best interest can perpetuate conflicts between populations. Good laws applying to everyone will gradually evolve.

Charity and insurance

Christianity teaches that we should be charitable, electing our wisest citizens to govern allotments. Advocates for particular sectors should communicate their wisdom, not create a protest group. Governments provide insurance cover against accident, illness and the consequences of ignorance. Insurance companies' interests discourage taking precautions. Better design can mitigate disasters.

Natural resources and real estate

Retail prices should incorporate the cost of researching best use of natural resources, purchasers are responsible for recycling. Real estate is valued by replacement cost, *e.g.* if its suitability for building isn't fully exploited, the highest bidder is free to develop it. Judicious taxation encourages resource conservation.

Money

An hour's work is a better exchange standard than gold, which changes value and the owner's fortune arbitrarily. Equating £10 with an hour's work, adjusted for overtime and social conditions, creates a stable, reproducible, international currency unit. Communal prosperity is best served by keeping everyone in productive work. If magnetic cards or similar devices are used, transactions can be monitored. With agreed limits on working hours and population size, the £10/hour standard will be stable.

Education

Adam and Eve were ashamed at their nudity when they ate the fruit of the *Tree of Knowledge*. We stand naked and unashamed before our teachers, revelling in learning. Everyone exchanges knowledge throughout life, as both teacher and student. Basing access to educational institutions on merit ensures best use of qualified teachers. Experience qualifies everyone to do some teaching.

Crime

Neighbourhood relationships and technology prevent property theft. Educating and rehabilitating offenders helps. Passion, revenge or mental illness leading to violent crime can be addressed by teaching relationship skills, resolving conflicts diplomatically and researching mental disorders. Sharing moral values, honouring legal agreements and offering advice reduces euthanasia and suicide.

Enterprise and tax

Patent and copyright laws protect innovations offering better ways to run society. Profits taxes have variable rates, that payable at the 90% rate is based on the logarithm of gross income. The tax on a profit of £1 B, is obtained by multiplying the tax rate by its logarithm, 9, yielding 90% x 9 = 8.1, anti-logarithm £125,900,000, see Table 3.1.

	Gross income			
	1,000,000,000	1,000,000	1,000	100
Tax rate	Income after tax			
90%	125,900,000	251,000	501	63
80%	15,850,000	63,100	251	40
70%	1,995,000	15,850	126	25
60%	251,200	3,981	63	16
50%	31,620	1,000	32	10

Table 3.1 Income after tax, man-hours

If people respect the greater good, they need fewer controls; increasing personal responsibility obviates bureaucracy. Taxes fund defence, education and research.

Leaders

The best respected leaders lead by example, execute power judiciously, aren't corrupt and consult the electorate; they regulate entrepreneurs' influence on society.

3.2 Reliance on truth

Learning, humanity's greatest strength, makes successive generations more prosperous. Isaac Newton said: *If I have seen further than others, then I have stood on the shoulders of giants.* Our forebears replaced religious creeds with scientific knowledge by devoting their lives to discovery. Explaining discoveries contradicting accepted dogma and having them acknowledged takes time. Applying the latest knowledge justifies actions.

Science, mathematics and art

Astronomers and physicists seeking an explanation for everything are rarely satisfied. Instrumental design limits precision and factors might have been overlooked. The language of mathematics condenses relationships applying to the entire universe. The future would be brighter if everyone recognized the truth. Artists examine society's soul, complementing cultural deficiencies.

Madness and genius

Each of the mind's nine independent, mathematically orthogonal parts sees the world differently. Deciding whether creative, analytic or artistic individuals are sane is difficult. Sanity and normality are relative; a clown attending a party sombrely dressed or a church minister visiting a nudist colony are both sane but acting out of character. The clown may have lost his mother or the minister preached about the Garden of Eden, their actions aren't abnormal.

Explaining new ideas is difficult if they contradict accepted dogma. Innovators may be ostracized and their views outlawed. Deviants have been burnt at the stake, visionaries confined to caves and cellars, lunatics exhibited in asylums, heretics interrogated and unorthodox philosophers censored.

Psychiatrists should educate, not isolate their patients, enhance their personalities and accept their eccentricities. They should encourage creativity and distinguish genius from lunacy. Science advances by challenging consensus beliefs, ostracism and psychiatric referral don't help.

Basic and absolute truths

Perhaps the *Tree of Knowledge* embraces all knowledge with just nine axioms and nine branches, explaining everything. Understanding them would solve all problems. Research nourishes the tree's roots, elucidating difficulties, new knowledge may require branches pruning.

Religions perpetuate their founders' insights, scientists seek absolute truth.

Privacy and secrecy

Concealing truths: failing to declare assets to customs or tax collectors, a teenager getting her boyfriend into trouble by withholding her age or a nation defending itself with secret weapons can be hazardous. Sharing knowledge leads to peaceful relationships and enhances trust. Society progresses by explaining and refining underlying principles and incorporating new ideas. Their discoverer's right to privacy must be respected.

3.3 Reliance on beauty

Beauty lies in the eye of the beholder, in the listener's ear or on the taster's tongue. Our physical senses judge creation's beauty and we surround ourselves with pleasing objects. Art or music appreciation groups share tastes and aesthetics when studying works by such classic artists as Van Gogh, Michelangelo or Monet or modern artists like Picasso or Warhol. Art provides an anchor against change, outliving technology, stabilizing and unifying society by linking generations.

Individual health, medicine and harmony

Lean, muscular bodies achieved through exercise and diet are beautiful. Sedentary lives with poor diets lead to obesity and other health problems. Exercising, hygiene and balanced diets maintain mental and physical health. Part 2 recommends a suitable diet. Beautiful environments also contribute; respecting other people's aesthetics makes for harmonious relationships.

Without state sponsored prevention, diagnosis and treatment, maintaining good health is expensive. Relatives, friends and neighbours may supplement state provisions. If terminal illness is painful, medical professionals can assist in ending life.

Beauty in communications and relationships

Expressing an opinion is difficult without a common language. Depending on their beauty, words may inspire either activity or indifference. An epic ballad can reduce a strong man to tears, ingenious advertisements encourage customers to prefer your products. Beauty isn't necessarily skin deep, but it's a weak basis for personal relationships. For compatibility, all the mind's nine coils are invoked, relationships can fail if knowledge and aesthetics aren't shared.

Pollution and the conservation of resources

Over-exploitation of resources and waste disposal have polluted the world. In many areas, drinking water is unsafe and acid air pollution addles minds. The ice caps are melting, ozone layer damaged and species endangered. We've disregarded natural beauty, reduced consumption, outlawed obsolescence; recycling could conserve it. Employing efficient solutions and conserving resources is paramount for our survival.

3.4 Striving for peace

John Lennon wrote IMAGINE:

Imagine there's no heaven
It's easy if you try
No hell below us
Above us only sky
Imagine all the people
Living for today...

You may say I'm a dreamer
But I'm not the only one
I hope someday you'll join us
And the world will be as one.

Imagine there's no countries
It isn't hard to do
Nothing to kill or die for
And no religion too
Imagine all the people
Living life in peace...

You may say I'm a dreamer
But I'm not the only one
I hope someday you'll join us
And the world will be as one.

Imagine no possessions
I wonder if you can
No need for greed or hunger
A brotherhood of man
Imagine all the people
Sharing all the world...

You may say I'm a dreamer
But I'm not the only one
I hope someday you'll join us
And the world will live as one.

We can realize ideas and wishes by employing body, mind and spirit. Medicine and technology care for our bodies, sharing knowledge relaxes our minds and focusing on intellectual development keeps our spirits up. Peace in our time, our greatest hope, may be achieved by seeking to be at peace with ourselves and uniting the human race. Atheism and pacifism may contribute positively to social welfare.

Competition and defeat

Man is competitive, striving to outshine his neighbour's garden, perform well at school or drive a better car. Competition doesn't necessarily leave winners happy if losers are jealous and resentful. Our greatest competitor lies within, we're our own worst enemies, upset if we fail to win. Our failings are innate, best overcome by learning from experience. Losing a competition shouldn't defeat us, we can learn to win internal and external peace.

Communications and travel

Mankind uses speech, prose, poetry, music, and art to communicate – pictures are worth a thousand words, sculptures a thousand conversations. Western people spend much time absorbing mass media instead of thinking or exercising free will. Restricting sources of information is risky, it needs sharing, reviewing and researching. World citizens travel for employment, education, trade and pleasure, showing a passport at national boundaries and respecting different cultures. Journeys may extend beyond the planet when life on Earth is secure.

Everlasting international peace

Scholars, diplomats and citizens following CIS, Compulsion for International Service could reconcile national governments, learn foreign languages and respect alien religious beliefs and ethics. To achieve international peace, everyone must participate. Conflict and discord have dominated history, battles been fought over national frontiers. Exchanging knowledge will diminish strife.

Peace of mind

Socrates said *'know thyself'* – try to understand your strengths and weaknesses on nine axes to achieve peace of mind. The Buddha may have succeeded, but I'm a doubting Thomas. Everyone 'bears a cross', has some handicap. Retreats can help, but the value of ritual prayers in churches, mosques and temples is limited. Your life may end with peace of mind if you study and practise the nine ways.

3.5 Striving for love

Most creatures' emotions are limited to happiness, anger and fear. Dogs exhibit loyalty to their owners and remorse when they offend them, elephants mourn their dead and dolphins seem to enjoy life. Human emotions can overwhelm us when we fall in love, get butterflies in the stomach or feel a glow of pride in an achievement.

Love is incomprehensible but comprehensive, boundless but binds like iron chains, obeys no laws, but is perfectible. Our desire to love and be loved may precipitate irrational acts, override our intentions or redirect our lives. Ideally, we love everyone, suppress hatred and avoid selfishness, jealousy and greed, avoiding Romeo and Juliet's tragedy.

Devotion and mind

We devote ourselves to and protect those we love, investing much time and energy nurturing relationships. Clinging exclusively to your partner can breed jealousy and restrict cooperation with others. You must reconcile devotion to the one you love with your free will, it helps to understand how minds work.

Procreation and permissiveness

In good times, reproduction expands opportunity: larger populations learn, progress and build more; in lean times, it preserves us from extinction. Our big heads mature slowly with a long stay in the womb and an extended childhood, humans are immature for longer than other species. Each generation may exceed their parents' achievements through education.

Everyone shares responsibility for educating, parenting and raising children to become productive, balanced adults, learning from children's untutored minds as they teach. Unless it threatens irreparable harm, we permit children to learn through experience, postponing some experiences until their consequences are understood.

Love of self and others

To achieve inner peace, love yourself and try to engage all the mind's nine facets in understanding others. Loving yourself becomes easier by studying all the nine ways, leading to inner peace. Self-hate alienates potential friends and creates strife with neighbours. Start by trying to love everyone, reconcile differences and reinforce relationships to achieve inner and outer peace.

Love, marriage and jealousy

The prime advantage of marriage is providing a secure environment for the children's benefit. Children raised under positive parental influences in a nuclear family fare better than in a broken home. Jealousy infringes peace, prevents sharing your mate's attention with anyone and restricts your partner's contributions to society. Clinging lovers hold each other back.

Grace

Communication connects the chain of love and peace. When we speak peacefully, peace breaks out. We're loved if we love others and treat everyone with grace and goodwill to achieve progress, the driving force of the New Age.

3.6 Striving for progress

Nationalism, religious fanaticism, racism, resource abuse, corruption and environmental degradation must be avoided to allow progress. Here are some pitfalls to avoid when trying to improve the world:

Necessity

Necessity is the mother of invention, inventors solve real, not imaginary problems.

Recycling

Resource availability limits progress, over-exploited natural resources need conservation. As populations grow, recycling becomes essential.

Integration

Novelty confuses some people; educators can help everyone to adapt, support inventors and ease the introduction of inventions.

Dedication

Panels of experts allot research funds to scientists dedicating their lives to discoveries which may yield no financial profit. Intellectual property rights acknowledge scientists' contributions.

Freedom of enquiry

Discoveries like my seminal finding at Cambridge come serendipitously to prepared minds, rarely by organized research. All my ideas stem from that chance discovery, contradicting scientific dogma, consensus beliefs and established principles. I hope to tell everyone about their consequences, rewriting the underlying rules and axioms of science accordingly. Doubts remain, debate and knowledge exchange leads to progress.

Intentions

Progress emerges from blue sky research, not by setting goals. New techniques and approaches can improve everyone's lives. A state of stability, justice and unity will prevail when the New Age ends, meanwhile, progress is essential.

3.7 Freedom from stability

As the New Age dawns, stability, justice, and unity are less important than peace, love and progress, but a stable currency, industrial harmony and shared moral values make way for change. The utopias promised by prophets can be sampled, allowing society to evolve.

Philosophy

Ruling philosophers seek ways to improve society, reviewing new ideas and promoting their exploitation. As Plato's *The Republic* suggested, selecting a philosophic leader is tricky, they need balanced personalities, be prepared to answer challenges to their philosophy, embrace and encourage change.

Peace and progress

Peace affords freedom to think, changes need periodic review.

External stability

Stability and having people we trust to solve our problems are comforting in times of need. Freedom to make changes enables us to help our neighbours be happy.

3.8 Freedom from justice

Deviant behaviour invokes a call for justice and law enforcement. We try to love everyone, rehabilitate, not seek revenge and retribution. Law enforcement should reflect the spirit rather than the letter of the law to preserve freedom. Officious law enforcement can breed opposition.

Law, relativity and spontaneity

Realistic laws need reflect normal social behaviour, be enforceable and consistent with social mores. Newton's laws of motion predict what happens when a ball is thrown or a bird takes flight. By allowing for the speed of light, Einstein's relativity correctly predicts the courses of planets and satellites. Newton and Einstein shared the language of mathematics.

Lawyers need a common language to debate and implement justice. Children's sense of justice is guileless and spontaneous, laws echoing truths learnt as children need no further justification.

Mutual understanding and religion

Reconciliation and compromise cause discontent, nobody puts all their cards on the table. Treaties need be comprehensive and embrace disparate opinions to avoid enmity. Copernicus was ridiculed for proposing the Earth orbits the Sun, it's now widely accepted. New truths are easier to adopt than new laws, which can lead to political or religious conflict.

Science challenges dogmatic religious beliefs, some religious sects endanger social cohesion, but their censorship can lead to martyrdom. Religious practices that are intolerant or irreconcilable with peace and justice need be outlawed. A secular society can embrace all followers of religions.

3.9 Freedom from unity

The unity created by religious, national or community groups affords spiritual comfort, military defence and better living conditions. Such organisations are divisive if they exclude outsiders.

Entropy, limitations, and censorship

Entropy is the scientific measure of disorder. When molten iron solidifies, becoming more ordered, it loses entropy. Unification should limit action, not thought, leaving individuals free to express their ideas without unreasonable censorship.

Unity of nations and democracy

As international treaties connect disparate peoples, differences across national boundaries will diminish. Universal standards will prevail over incompatible regimes and currencies. Winston Churchill said: *Democracy is the worst form of government, except for all the others.*

World government can evolve gradually by allowing leaders like Barack Obama and Nelson Mandela and organizations like the United Nations, *Médecins Sans Frontières* and Oxfam to prevail and encouraging participation in VSO and Peace Corps. Decentralizing government allows local reinterpretation of central policies and dispute resolution. UN surveillance of national elections and international diplomacy help eliminate dictatorships. Appointing the wisest individuals to unelected bodies like the UK's House of Lords enables them to promote common welfare and evaluate new political ideas, discoveries and inventions, ensuring they'll benefits everyone.

Toward a unified new age, totalitarianism and mind control

Unity arising by default permits everyone the freedom to pursue peace. As goodness, truth, and beauty are taken for granted, we'll strive for peace, love, and progress free from stability, justice and unity.

Recent history includes many examples of democracy replacing totalitarian regimes, dictators being overthrown and purges, censorship and indoctrination being eliminated. The use of drugs, indoctrination and torture has diminished. This is the dawning of the New Age when our minds will be free and at peace and the problems raised in this book will diminish.

Conclusion: CIS

The prefix *cis-* comes from Latin, meaning together or on the same side, I propose CIS, standing for Compulsion for International Service to embody the New Age. For its realization, we must broadcast the idea globally, visit and connect with foreign groups and help our neighbours. Connections established by offering our services abroad transcend national governments. Sharing knowledge and understanding opens borders and breaks down international barriers.

In the New Age, the Age of Man, mankind will progress with open minds and mutual help. Those reluctant to abandon goodness, truth and beauty may object to peace, love and progress. Some may doubt change is possible or necessary, and see the New Age as a pipe dream. I address those wishing to pursue new ideals, change and progress and seeking to secure new freedoms for everyone, asking '*How can we bring it about?*'

Part 2: What's natural?

Preface

People often distinguish between natural and artificial, treating products of man's intervention as inferior or unnatural. This book seeks to describe a propitious environment for human survival. Some resources are exhausted by the expanding population; polluted air, land and seas have arisen as lifestyles evolve. New ways of doing things render old ways obsolete. Corruption assaults the mind and senses. I'm concerned with the lives of human beings in sickness and health and how they can lead balanced lives in a complex society.

4 Molecular control

Many students regard chemistry as complicated. Enzymes and chromosomes are large, complex molecules, even those of sugar can be confusing. We use cars, cell-phones and computers, understanding their purpose and operation but not necessarily how they work. Likewise, the purposes of our body tissues, cells and the interactions of the molecules they contain can be described without understanding all the details. It's the underlying principles which are important.

Formulae and equations

Biochemists use many of the longest words in English, chemical formulae can be complicated, sketches using symbols and abbreviations help. Atomic elements are represented by capital letters, some followed by a lower-case letter with subscripts connoting their number. *E.g.* H_2O represents a water molecule having two hydrogen atoms, H, and one oxygen, O. Adenosine triphosphate, ATP, $C_{10}H_6N_5O_{13}P_3$, contains 37: 10 carbon, 6 hydrogen, 5 nitrogen, 13 oxygen and 3 phosphorus atoms. Each has stable 4-way, tetrahedral connections, they don't jostle together at random, Fig 4.1.

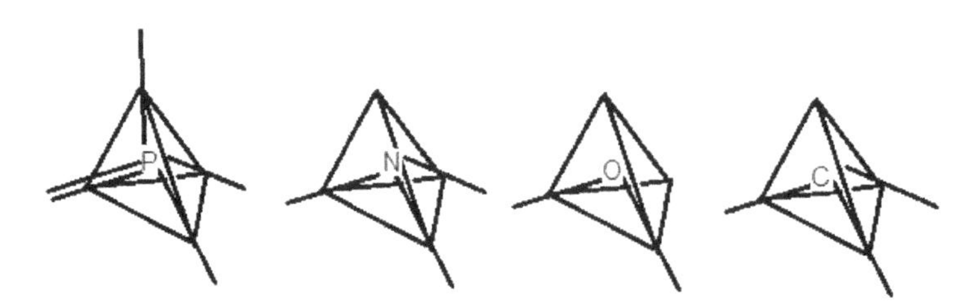

Fig 4.1 P, N, O & C atoms are tetrahedral

Each element has a characteristic number of connections to others, 4 for C, 3 for N, 2 for O and 5 for P; hydrogen bonds, H-bonds are represented by dashes -----, single bonds by lines —, and double bonds by double lines ===. When C, N or O make double bonds, the structure is triangular, Fig 4.2:

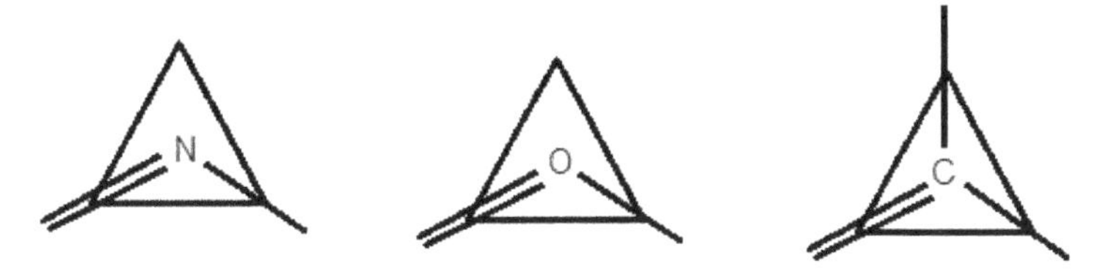

Fig 4.2 N, O & C atoms can be double bonded

Loose ends signify H's. The slightly negatively charged, electronegative N, O and P bond with the slightly positively charged, electropositive H, connected by directed, dipolar, proton-ordered H-bonds determining the shapes of proteins and chromosomes. Chemical equations such as:

$$O^{=} + 2H^{+} \rightarrow H_2O$$

must balance, with the same number of atoms on each side. The superscript = in this equation connotes the double negative charge on oxygen, neutralized by the positive charges on protons, H^+. Single bonds connect O and H atoms, H-bonds form between adjacent water molecules. H-bonds make and break easily; when those between O and H are exposed to four micrometre, 4μ wavelength infrared light, they convert chemical to mechanical energy.

Later sections describe processes such as muscle contraction, sperm swimming and cell division in which they participate. Most chemical reactions involve making and breaking strong single and double 'covalent' bonds. Because molecules move at about the speed of sound, chemical reactions are too fast to observe. Their behaviour resembles an unruly crowd of people, obeying the statistical laws of thermodynamics *en masse*.

Structure of ATP

ATP has three parts: adenine, **a** ribose, **b** and triphosphate, **c** and combined, **d** with a shorthand version for A, □ at **e**. Lines implying C and H, make it easier to draw and understand. The shorthand for Thymine, Guanosine and Cytosine is +, Δ and *, replacing T, G and C. Two H-bonds connect bisymmetric □+ and three tri-symmetric Δ*, Fig 4.3.

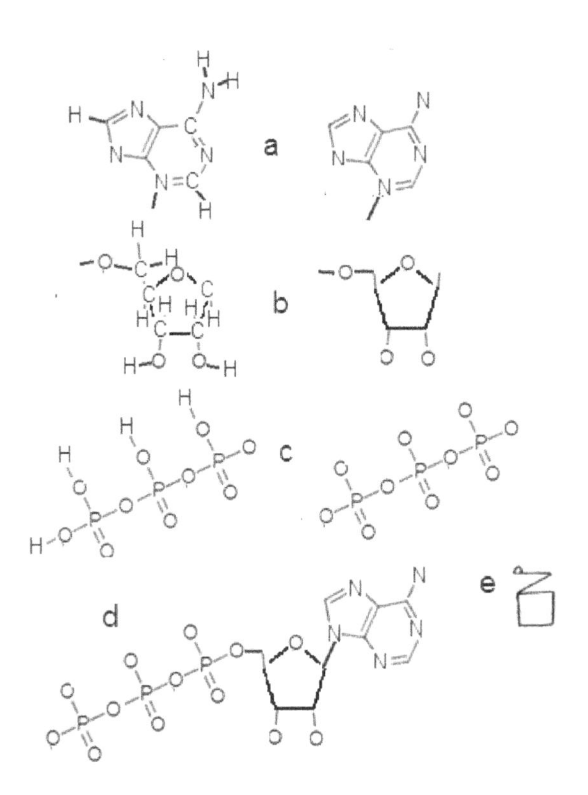

Fig 4.3 Structure of ATP

Concentrations

The body is mostly composed of water, breaking polarized solute molecules into charged ions. The OH groups on ATP become hydroxyl,

OH^- ions, its triphosphate can release a phosphate, PO_4^{3-}, P_i or diphosphate, $P_2O_7^{=}$, PPi ion. Square brackets, [ ] connote concentrations and pH a cell's acidity, pH is defined: $pH = - \log_{10}[H^+]$, where $[H^+]$ means H^+ concentration. I argue in 5.6 that acid air pollution by sulphur or nitrogen oxides, SO_x or NO_x changes the pH of cells at the back of the nose, causing Alzheimer's disease.

The sugar concentration after adding 1- 10% sugar to a drink is 10^{-100} M ppb, parts per billion. Adding an ATP molecule weighing about 10^{-21} g, $^1/_{1,000,000,000,000,000,000,000}$ grams to a cell whose volume is 1000 μ^3, one thousand cubic micrometres, makes an ATP concentration of about 0.1 ppb. Numbers of molecules per cell are a better measure of such miniscule amounts. Accidentally dropping a grain of sugar into a bucket of water makes a concentration around 10 ppb.

What's natural?

Supermarket products are often labelled 'natural' or 'using natural ingredients', 'natural' is regarded as superior to 'artificial' and human intervention as 'unnatural'. All raw materials are natural and we're natural products of evolution. Conservation considerations are the best criteria for choosing the best sources, 'green' manufacturers take the environment and waste disposal into account.

This book explores ways to balance nature's dynamic equilibrium and human welfare with ecology. Competitive marketing exploits naive customers by using such corrupt practices as built-in obsolescence, cosmetic improvements and incompatible components. Conserving natural resources becomes increasingly challenging as the human populace expands.

Education, leading to smaller families, could resolve this dilemma. Our society of naked apes need draft a new educational curriculum. Molecular societies obey the laws of science whilst human societies are self-governing. Legislatures taking scientific advice will enact rules of conduct compatible with nature.

4.1 Molecular pumps

Cells are surrounded by double membranes, walls of opposed fatty acid molecules, creating a detergent barrier to maintain the concentrations of their contents and exclude undesirables. Molecular pumps, transport DNAs, tDNAs made of nucleic acid, feed controlled quantities of nutrients through pores, Fig 4.4:

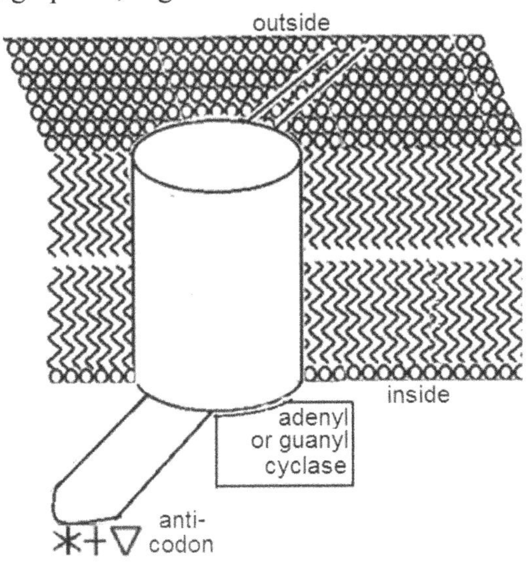

Fig 4.4 tRNA or tDNA pump in membrane

The fifth molecule shown in Fig 4.5 is methyl adenosine monophosphate, AMP, with extra methyl groups. Methylated nucleotides form the fine details of molecular pumps; I call them *fancy* nucleic acids.

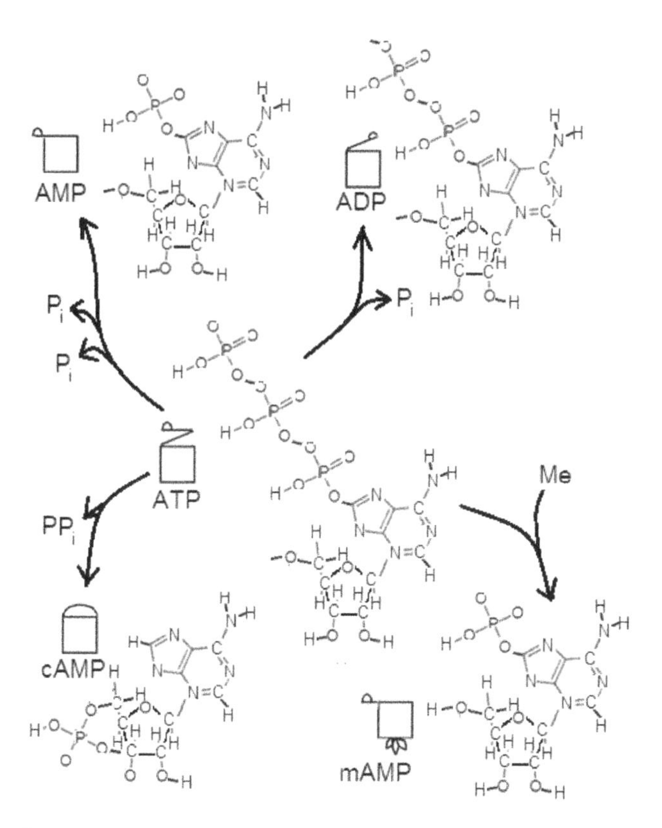

Fig 4.5 Molecules related to adenine phosphate

Batteries

ATP serves as a battery storing energy released by breakdown of sugar. Energy is released when ATP breaks down to adenosine monophosphate, adenosine diphosphate or cyclic adenosine monophosphate, AMP, ADP or cAMP and phosphate or pyrophosphate, P_i or PP_i; Fig 4.5 shows their shorthand representations:

These molecules all incorporate adenosine, □. Its triphosphate, ATP stores energy for all purposes; the reactions:

$ATP \rightarrow ADP + P_i + \sim\rightarrow$ and

$ATP \rightarrow cAMP + PP_i + \sim\rightarrow$

release energy 4μ energy, ~⟶ for most operations, the first in the citric acid or Kreb's cycle in mitochondria or photo-phosphorylation in plants' grana, the second drives active transport. ATP is recharged by phosphorylation.

Nucleic Acids

There are two types of nucleic acids: ribonucleic acids, and deoxy-ribonucleic acids, RNA and DNA. DNA gets its name from the O missing from its ribose sugar. Both have important roles in transporting cell nutrients, the outer cell membrane uses DNA pumps. Membranes of such sub-cellular components as mitochondria and grana use RNA pumps. RNA is composed of four bases: Adenosine, Uridine, Guanosine and Cytosine, A, U, G and C connected by ribose. Thymidine, T replaces uridine in DNA, see part 4. Fig 4.6 shows eight molecules containing these bases and their shorthand representations. Ribo-nucleic acids form chains of mRNA. □ paired with + and Δ with * form double-stranded DNA, Fig 4.7.

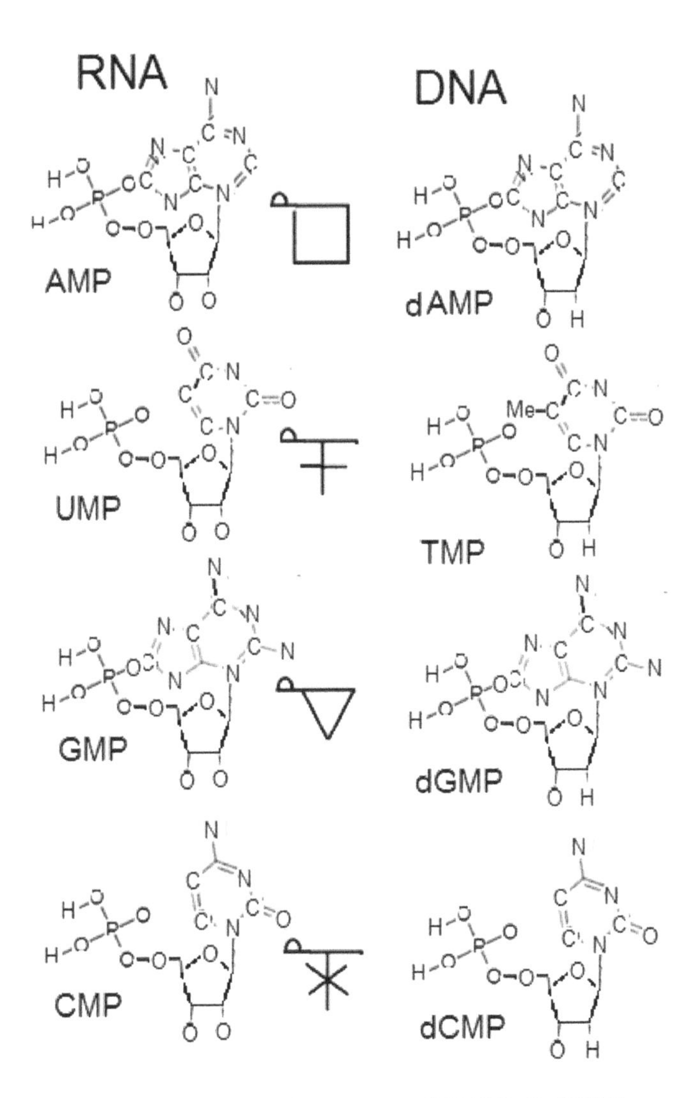

Fig 4.6 Constituents of RNA & DNA

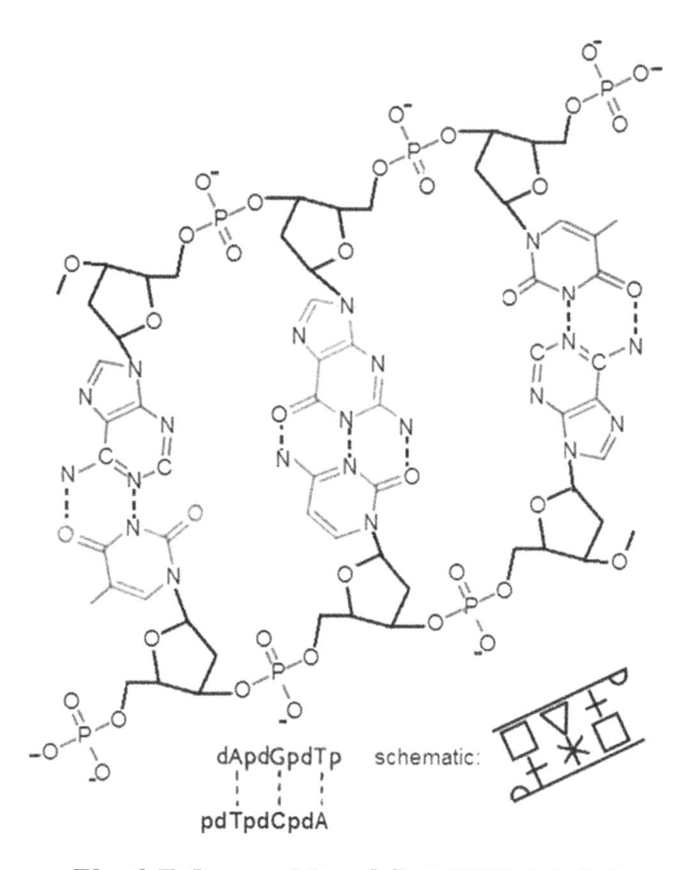

Fig 4.7 Assembly of flat DNA triplet

Pumps

The difference between transfer RNA, tRNA and transport DNA, tDNA pumps is their cargo. The pump wraps itself up, the 76 components of phenylalanine transfer ribonucleic acid, $tRNA_{Phe}$ from yeast are laid flat, **a** tRNA includes some fancy mAMPs. The arrows, **b** indicate how it bends, wrapping itself up through **c** and **d**, the lower loop forms a double helix. The corners are rounded**, e** leaving a complete pump, **f** shown in stereo, **g** and rotated to embed in the cell membrane, **h** and **i**, Fig 4.8, also see Fig 4.4.

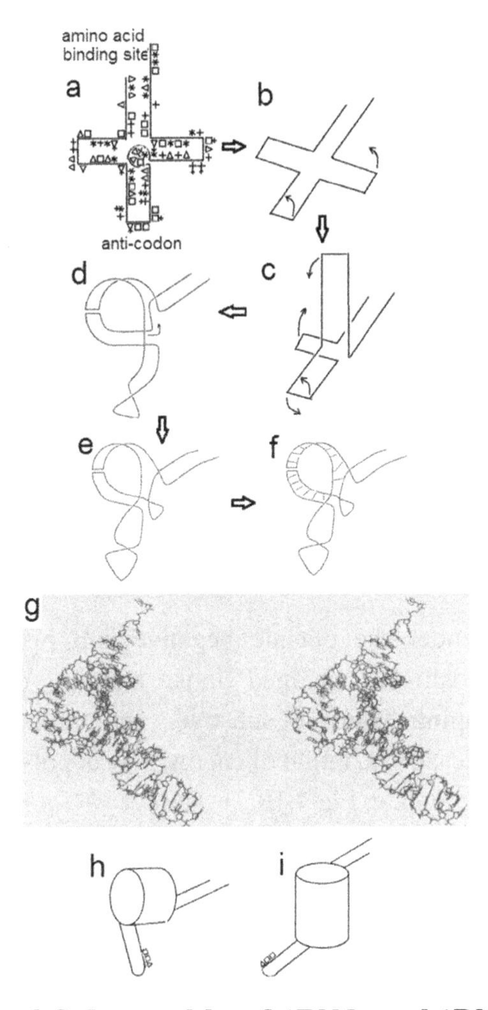

Fig 4.8 Assembly of tRNA and tDNA

The anti-codon triplet, ΔΔ□ pairs with complementary **+ on messenger RNA, mRNA or differentiation DNA, dDNA. Adenyl-cyclase drives active transport pumps, concentrating substrates, Fig 4.9.

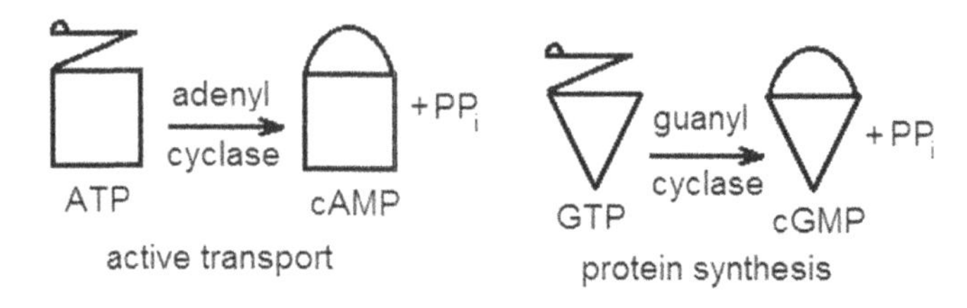

Fig 4.9 Adenyl/guanyl cyclase breakdown ATP/GTP

$$ATP \rightarrow cAMP + PPi + \sim\rightarrow$$

Guanyl-cyclase drives protein synthesis, adding amino acids to a nascent protein:

$$GTP \rightarrow cGMP + PPi + \sim\rightarrow.$$

Some pumps render the outside negative and inside positive by accumulating positively charged ions, cations, determining the direction of pumping. Transport substrates wait, **a** until adenylcyclase is triggered to release a quantum of energy, ~→ depolarizing the pump, **b** and passing through, **c**, Fig 4.10.

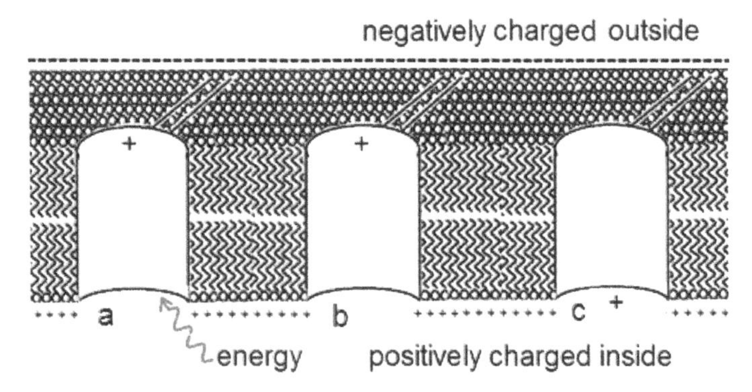

Fig 4.10 Pump mechanism

4.2 Genetic code

Chromosomes carry instructions for making messenger RNA, mRNA selecting transfer RNAs, tRNAs for protein synthesis and differentiation DNA, dDNA selecting transport DNAs, tDNAs which determine cell diet. The self-assembly of the double membrane, its fat molecules tail to tail with their heads making the membrane impermeable, is taken for granted.

Fig 4.4 shows non-functional molecular pumps floating freely. They're activated when triplets on mRNA or dDNA pair with complementary triplets, Fig 4.11; the first base can match in four ways: □, +, Δ, or * with +, □, * or Δ, the second and third each have four more ways to pair, making 4*4*4 = 64 triplet codes, *e.g.* tRNA_{Phe}'s ΔΔ□ with **+, encodes the amino acid phenylalanine, Phe.

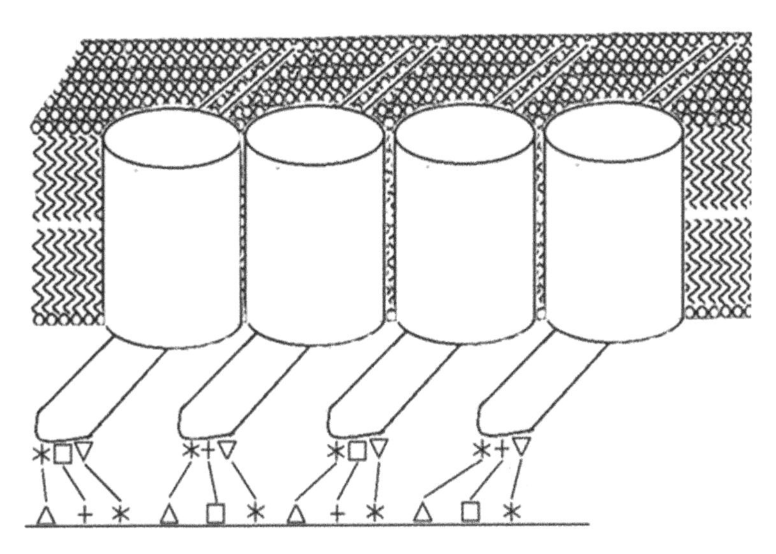

Fig 4.11 mRNA/dDNA bind adjacent tRNAs/tDNAs

The amino acids corresponding to the 64 tRNAs for protein synthesis are well established, Fig 4.12. Fig 4.13 shows a score chart for the substrates corresponding to the family of tDNAs when they're known.

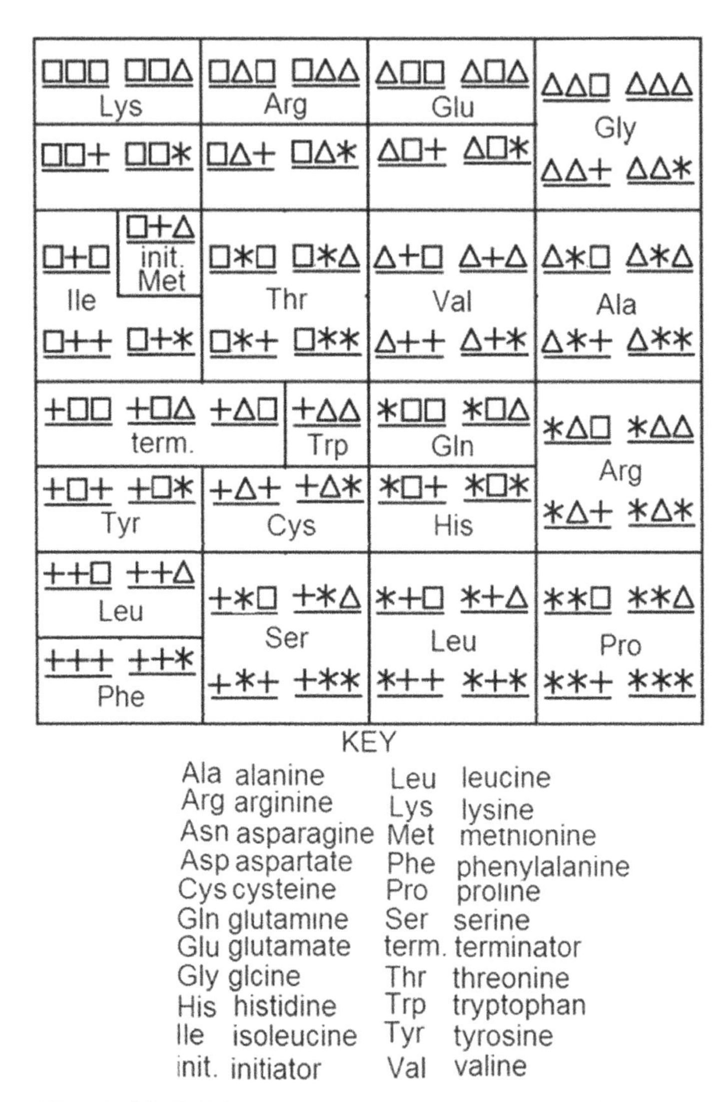

Fig 4.12 RNA triplet protein synthesis code

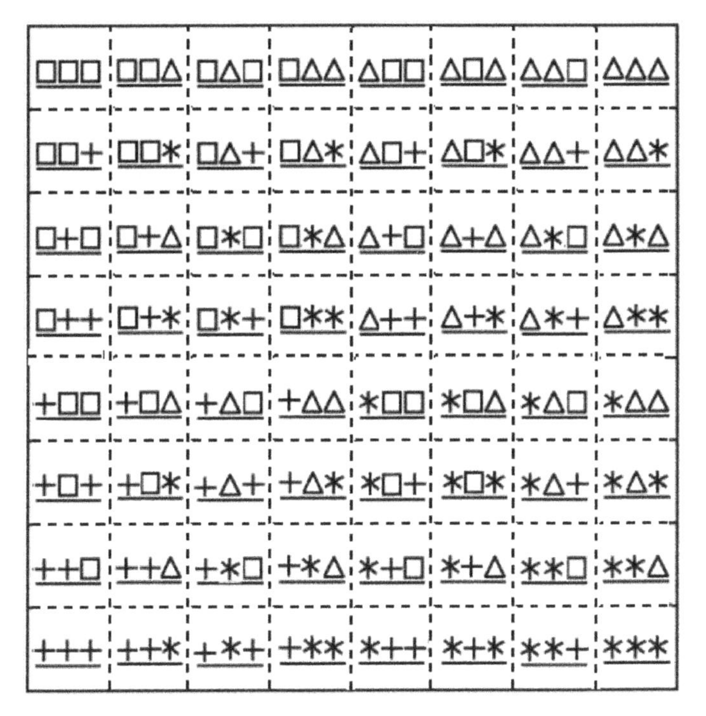

Fig 4.13 DNA triplet differentiation code score sheet

mRNAs are templates for protein synthesis on ribosomes, coordinating tRNA pumps, a mRNA typically encodes ~100 amino acids per protein, Figs 4.14 and 4.15.

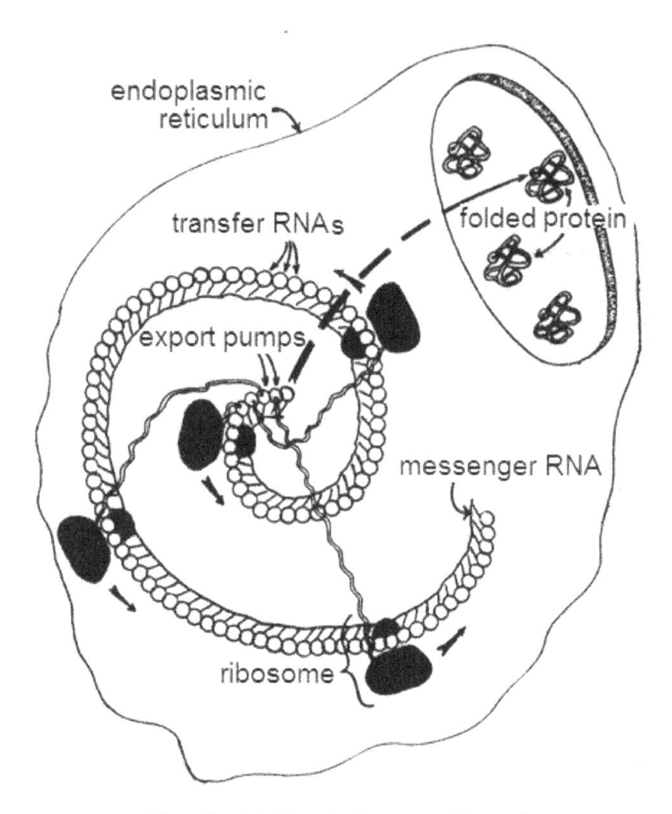

Fig 4.14 Protein synthesis

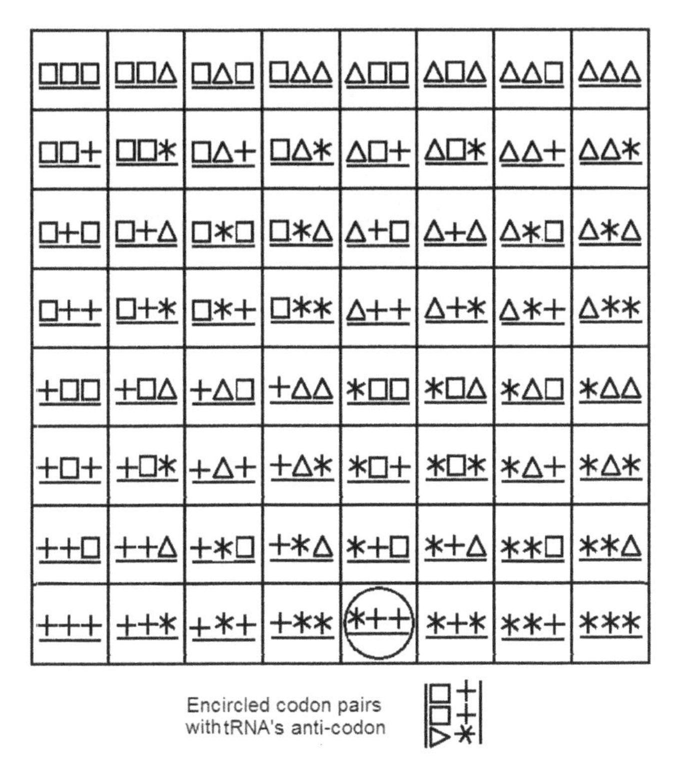

□□□	□□Δ	□Δ□	□ΔΔ	Δ□□	Δ□Δ	ΔΔ□	ΔΔΔ
□□+	□□*	□Δ+	□Δ*	Δ□+	Δ□*	ΔΔ+	ΔΔ*
□+□	□+Δ	□*□	□*Δ	Δ+□	Δ+Δ	Δ*□	Δ*Δ
□++	□+*	□*+	□**	Δ++	Δ+*	Δ*+	Δ**
+□□	+□Δ	+Δ□	+ΔΔ	*□□	*□Δ	*Δ□	*ΔΔ
+□+	+□*	+Δ+	+Δ*	*□+	*□*	*Δ+	*Δ*
++□	++Δ	+*□	+*Δ	*+□	*+Δ	**□	**Δ
+++	++*	+*+	+**	*++	*+*	**+	***

Fig 4.15 Detail of tRNA binding to mRNA

This chapter is primarily concerned with tDNA pumps encoding cell nutrition. 4.5 discusses how 'stem cells' differentiate to form body tissues by changing their diets by reversing a histone protein on a redundant dDNA, enabling the bound tDNAs to encode the synthesis of 'hook' proteins binding tissue cells together.

4.3 Carriers and hormones

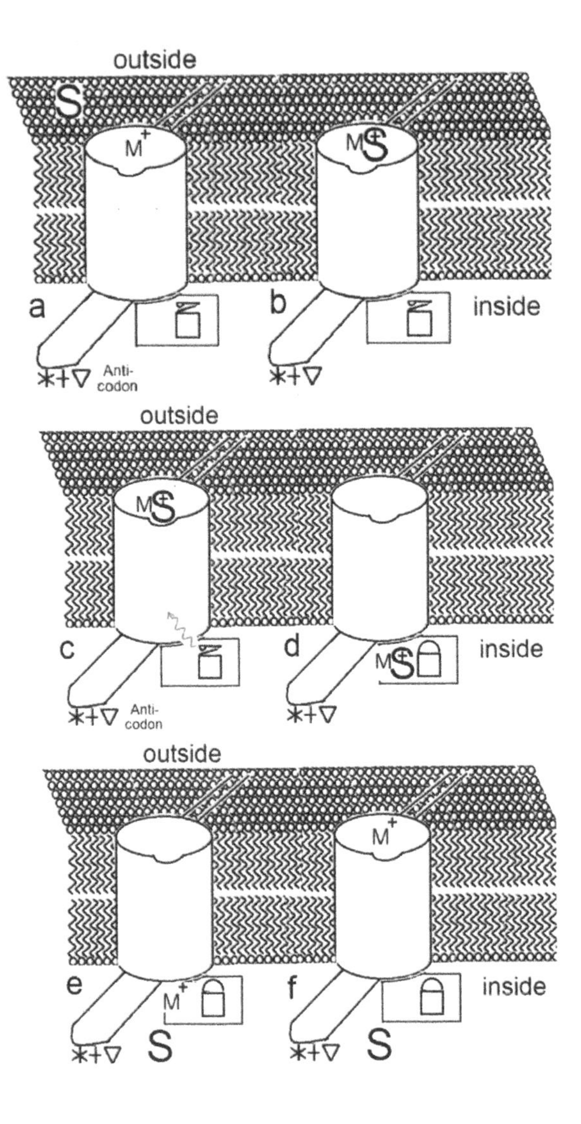

Fig 4.16 Charged carrier for uncharged substrate

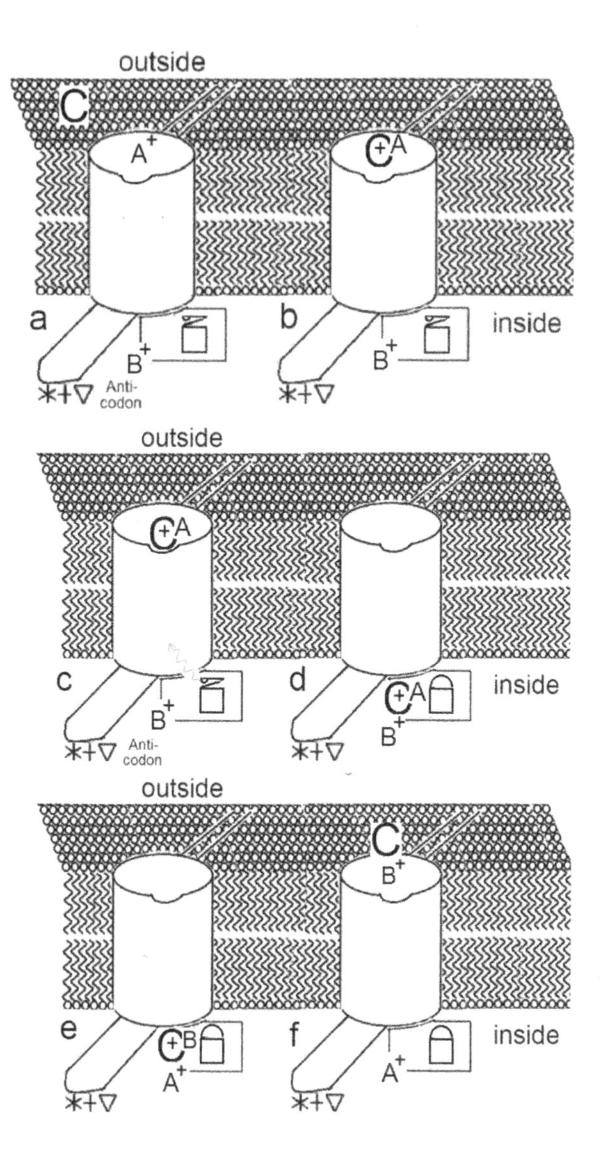

Fig 4.17 Uncharged carrier for charged substrates

tDNAs need priming before they can transport substrates into a cell, each has a matching carrier. Most pumps are primed by hormones delivering carriers.

Carriers

Fig 4.16 **a** and **b** show how an uncharged substrate, S binds to a positively-charged carrier, M^+ at the pump entrance, triggering ATP breakdown, **c** releasing energy to transport the complex, **d** into the cell. Inside the cell the substrate is released, **e** and the carrier returns to await another substrate, **f**.

In Fig 4.17 **a**, carrier, C binds a positively-charged ion, A^+ at the pump entrance, creating a complex, **b** triggering ATP breakdown, **c** releasing energy to transport it, **d** for exchange with another positively charged ion. Inside the cell the ion A^+ is released and replaced with B^+, **e** which is passed to the outer surface, **f**.

tDNAs using uncharged carriers exchange charged ions to change cell charge, *e.g.* adrenaline exchanges three potassium ions for two sodium, see 5.1.

Hormones

The *endocrine* glands use more tDNAs to import ingredients for making hormones. They're released as required, delivering carriers to tDNA pumps. In Fig 4.18 **a,** a hormone recognizes a tDNA, releases the carrier, **b** primes the pump, **c** and departs, **d**. Hormones have many roles: most deliver and remove carriers to start and stop pumps; others modify pumps and some play no part in active transport.

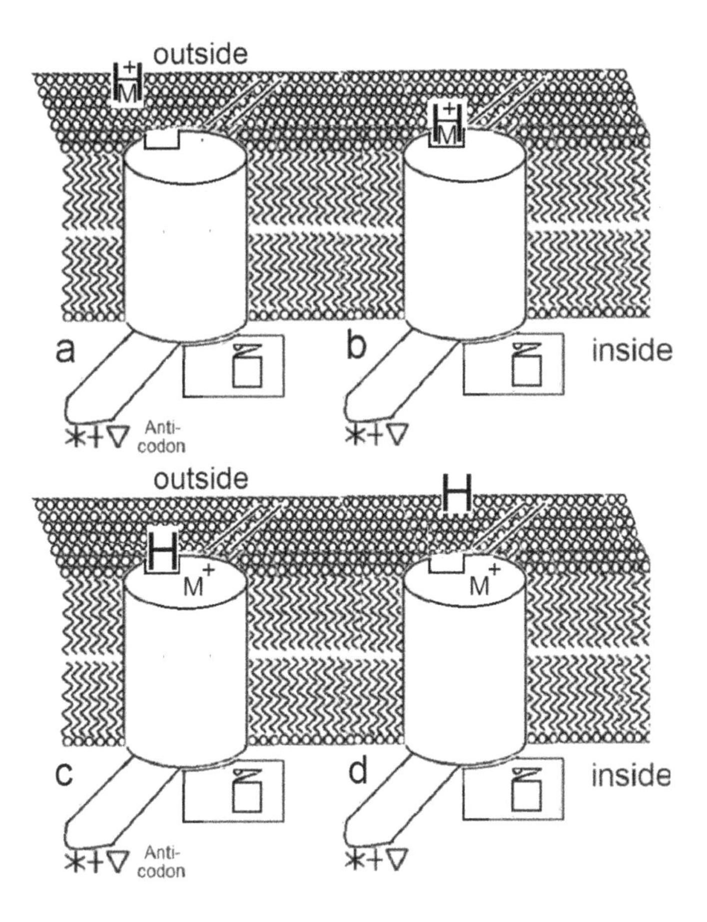

Fig 4.18 Hormone delivers carrier

Larger molecules

To transport nucleic acid and protein polymers, the carrier attaches to the first segment of the molecule and takes it in, then returns and reattaches repeatedly until the whole chain has entered the cell, Fig 4.19.

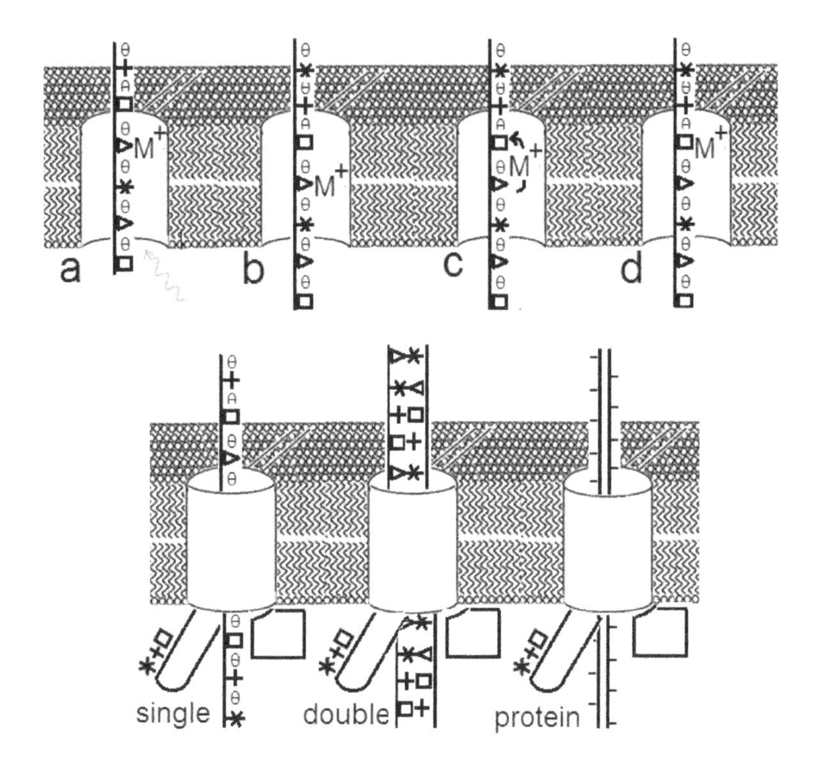

Fig 4.19 Polymer transport

Rhinoviruses causing colds and flu invade cells via un-primed tDNAs. Zinc, carrier for glucose, blocks vulnerable pumps in nasal passages, explaining the efficacy of zinc and its partner, vitamin C in preventing infection, see section 5.7.

4.4 Differentiation

Comparisons are usually between things in the same category, *e.g.* choosing a house from a range of dwellings. Individual's identities are influenced by nature, nurture and nativity – their parents' genetics, culture and upbringing and astrology, time of birth. Comparisons can be contentious.

Genetic deviation

Genetics determines an individual's fitness for survival. Genes encode intelligence, size, and body type, giving some the opportunity to live long enough to reproduce. Everyone has some *cross to bear*, but not all abnormalities are disadvantageous – differences can be beneficial.

Deviation at the cellular level arises through differentiation, coded by strings of □, Δ, + and, *. dDNAs select from 64 tDNAs using the same genetic language as mRNAs use to select from 64 tRNAs. Studying individuals with mutant tDNAs provides valuable insights into how life works.

Cell dinner

Genetic differentiation may be illustrated by substituting three-letter words for DNA triplets of □, Δ, +, and *, using nouns for cell diet and verbs for enzymes:

A DNA AIR GAS HAM HAM EGG EGG FAT END	B RNA CUT CUT FRY EAT END
C DNA JAM FAT BUN TEA END	D RNA CUT EAT EAT SIP END
E DNA EGG PIE ICE GIN END	F RNA EAT MIX SIP SIP END

Each gene, A to F, begins with RNA or DNA indicating whether it's an mRNA verb encoding protein synthesis or a dDNA noun specifying cell diet; all finish with END. For breakfast, gene A instructs tDNA pumps to ingest air, gas, ham x 2, egg x 2 and fat and gene B instructs tRNA pumps to make proteins to cut the ham, burn the gas-air mixture, to fry the ham and eggs, and eat them. Genes C and D make tea and E and F provide a nightcap of egg and pie with iced gin.

With all six genes active, the cell has breakfast, tea, and nightcap. If differentiated, it may be reduced to one or two meals. Without genes C and E it cannot eat tea or nightcap and only gets breakfast. With A and E blocked, diet is limited to tea; switching off A and C to nightcap. If B, D, or F are inactive, A, C or E need deactivating lest the cell gets indigestion.

Gene control

The histone protein beds, βββ-βββ-βββ, described in chapter 1 control genes. They hold the double-stranded nuclear DNA flat and confer directionality – copying always starts from the open end of the hairpin:

A			B	
DNA AIR GAS	HAM HAM EGG	EGG FAT END	RNA CUT CUT	FRY EAT END
βββ-βββ-βββ	βββ-βββ-βββ	βββ-βββ-βββ	βββ-βββ-βββ	βββ-βββ-βββ
C		D		
DNA JAM FAT	BUN TEA END	RNA CUT EAT	EAT SIP END	
βββ-βββ-βββ	βββ-βββ-βββ	βββ-βββ-βββ	βββ-βββ-βββ	
E		F		
DNA EGG PIE	ICE GIN END	RNA EAT MIX	SIP SIP END	
βββ-βββ-βββ	βββ-βββ-βββ	βββ-βββ-βββ	βββ-βββ-βββ	

With the hairpins on genes C and E turned round, δδδ-δδδ-δδδ, only gene A is readable, D and F can't be executed:

A			B	
DNA AIR GAS	HAM HAM EGG	EGG FAT END	RNA CUT CUT	FRY EAT END
βββ-βββ-βββ	βββ-βββ-βββ	βββ-βββ-βββ	βββ-βββ-βββ	βββ-βββ-βββ
C		D		
DNA JAM FAT	BUN TEA END	RNA CUT EAT	EAT SIP END	
δδδ-δδδ-δδδ	βββ-βββ-βββ	βββ-βββ-βββ	βββ-βββ-βββ	
E		F		
[DNA EGG PIE	ICE GIN END]	RNA EAT MIX	SIP SIP END	
δδδ-δδδ-δδδ	βββ-βββ-βββ	βββ-βββ-βββ	βββ-βββ-βββ	

When cells reproduce their chromosomes are copied, if histone hairpins are reversed, the daughter cells are differentiated; after a few generations of division cells become diverse. The effects of minor changes over three generations are shown here:

<table>
<tr><td colspan="8">βββ-βββ
ABC DEF</td></tr>
<tr><td colspan="4">βββ-βββ
ABC DEF</td><td colspan="4">βββ-βδβ
ABC DEF</td></tr>
<tr><td colspan="2">ββδ-βββ
ABC DEF</td><td colspan="2">ββδ-βββ
ABC DEF</td><td colspan="2">βββ-βδβ
ABC DEF</td><td colspan="2">βββ-βδβ
ABC DEF</td></tr>
<tr><td>ββδ-βδβ
ABC DEF</td><td>ββδ-βββ
ABC DEF</td><td>δβδ-βββ
ABC DEF</td><td>δβδ-βββ
ABC DEF</td><td>δββ-βδβ
ABC DEF</td><td>δββ-βδβ
ABC DEF</td><td>δβδ-βδβ
ABC DEF</td><td>δβδ-βδβ
ABC DEF</td></tr>
<tr><td>1</td><td>2</td><td>3</td><td>4</td><td>5</td><td>6</td><td>7</td><td>8</td></tr>
</table>

Offspring inherit mutations, cells progress from all having breakfast, tea, and nightcaps to cells 7 and 8 eating nothing. dDNAs with histones reversed make protein, see Fig 1.8. Palindromic pairs bind six base pairs uniquely, Fig 4.20.

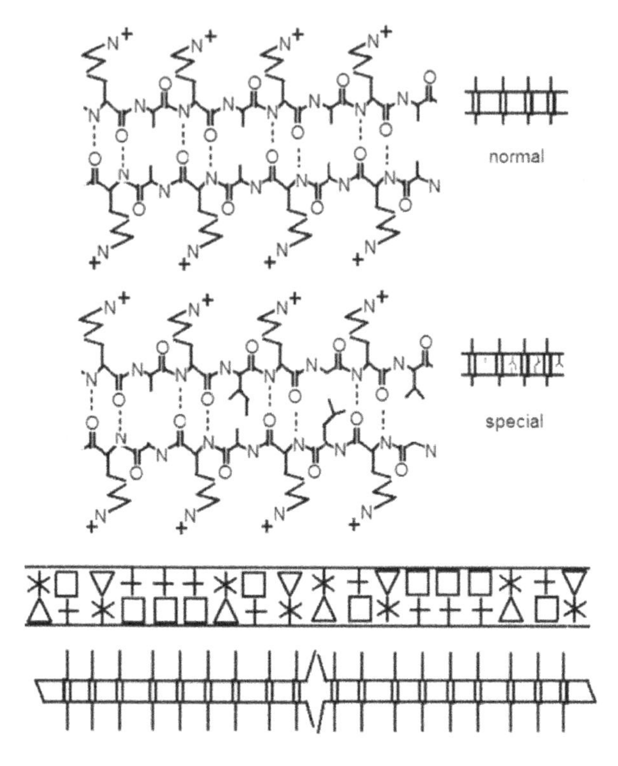

Fig 4.20 Protein β-sheet

By default, histones have alternate alanine and lysine, Ala and Lys. Replacing some Ala by isoleucine, valine or leucine, Ileu, Val or Leu, specifies the DNA sequence they bind to. A mutated dDNA might replace eggs with buns for breakfast.

My assumption that tDNAs are copied along with the chromosomes at cell division or form part of a bacterial genome is unverified, Fig 4.21.

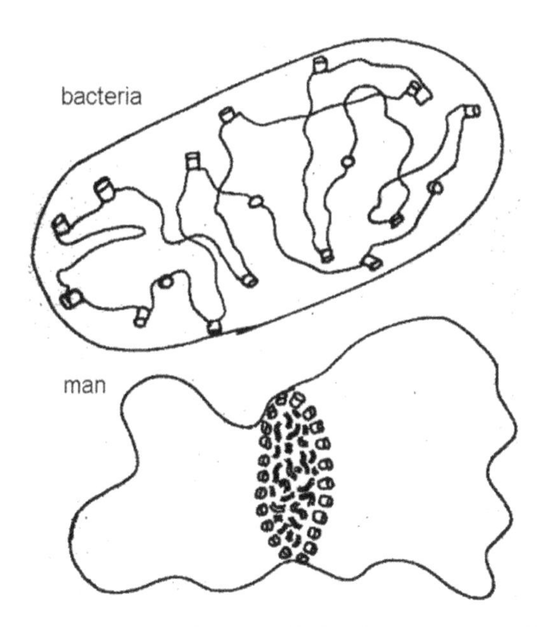

Fig 4.21 Human and bacterial tDNAs at cell division

4.5 Feedback

Protein synthesis on histones

Most protein synthesis uses tRNAs to translate mRNAs, tDNAs and dDNAs can also direct protein synthesis. Histone orientation determines whether adenyl- or guanyl-cyclase binds. With dDNAs binding tDNAs and adenylcyclase nutrients are imported; binding guanylcyclase, amino acids, ααs are concatenated, building a protein, Fig 4.22:

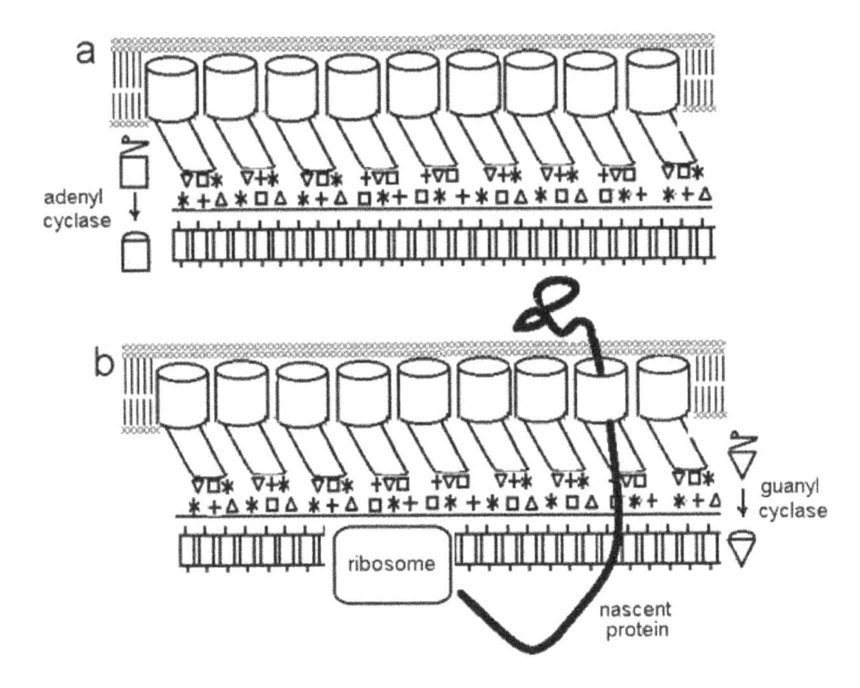

Fig 4.22 Adenyl-/guanyl-cylase power substrate/ amino acid pumps respectively

A tDNA feeds it through the membrane and anchors it, creating a *hook* connecting differentiated tissue cells together.

Tissue differentiation

Histone reversal arises when a string of active tDNA pumps runs dry and overheats due to ATP breakdown with nothing to pump. Cell differentiation forming different tissues accounts for embryology and metamorphosis, Fig 4.23:

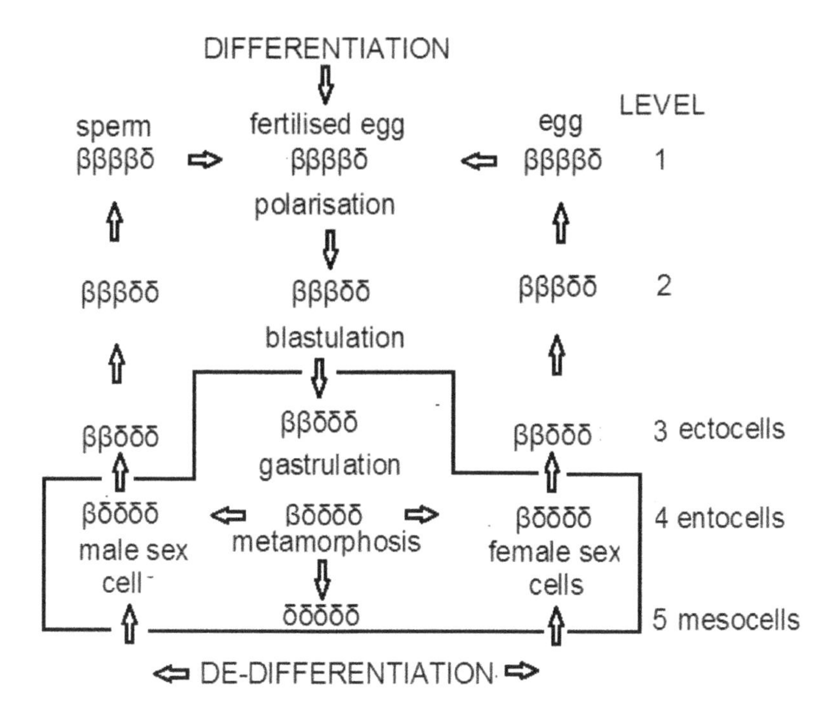

Fig 4.23 Differentiation versus dedifferentiation

Growth

The protein hooks pair with those on adjacent cells, binding them together. A cell with one hook attaches to one other cell, *e.g.* a sperm mating with an egg; cells with two hooks form a chain, *e.g.* algal spirogyra. Cells with three hooks form sheets, *e.g.* sponges; with four they create the double sheets, *e.g.* primitive worms.

After differentiating five times, cells have five hooks and can create all tissue types when combined with cells with fewer hooks. Cells with six hooks divide freely, resulting in cancers and tumours, Fig 4.24:

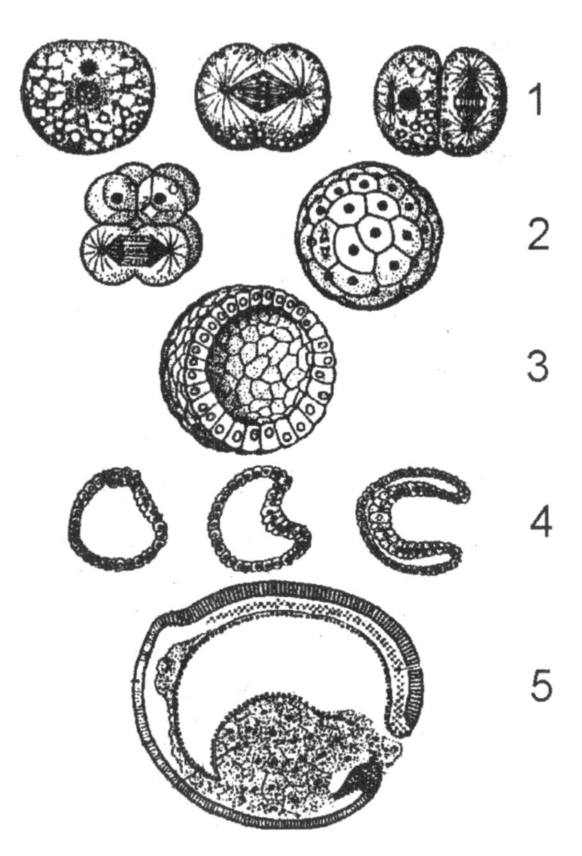

Fig 4.24 Five stages in tissue differentiation

Feedback failure also causes arthritis, obesity and allergies. Orderly differentiation creates the symmetries of plant flowers and our five fingered hands.

Embryology

Sperm cells swim to fertilize an egg cell, transferring DNA when their hooks engage. The resulting foetus then differentiates, forming another hook attaching it to the uterus. That first *stem cell* generates all types of tissue by dividing and differentiating as the child develops.

The first cell division creates bilateral symmetry; the second creates a hollow sphere of cells, the blastula. Its inward facing cells are starved of nutrients, differentiate and the sphere collapses. Another cavity forms, the gastrula; its starved inward facing cells also differentiate. When the three layers of cells surrounding this cavity differentiate again, they become:

1 Guts and other internal organs
2 Muscles, skeleton, heart and blood vessels and
3 The nervous system, brain and skin.

Dedifferentiation and haematopoiesis

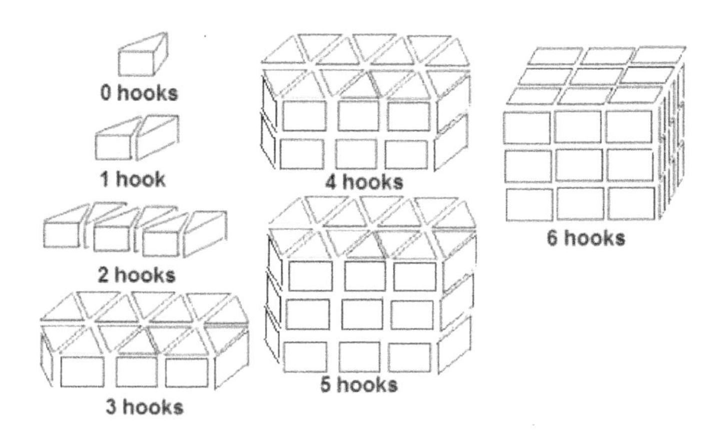

Fig 4.25 Five cell combinations, 6th allows tumour/ cancer formation

Differentiation running in reverse forms eggs and sperm; it has five stages like differentiation, a 'hook' is lost at each, Fig 4.25.

Blood cell creation, haematopoiesis in bone marrow, also takes five steps. Red cells, erythrocytes have masked hooks, preventing their sticking to one another. Leucocytes, a type of white blood cell, have exposed hooks which can pair with the sixth hooks on cancer and tumour cells or foreign bodies – the immune response, Fig 4.26:

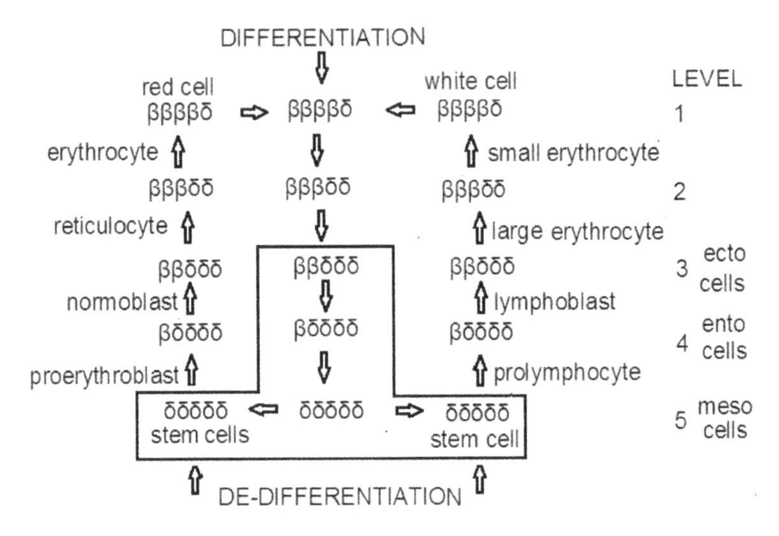

Fig 4.26 Four stages in blood cell differentiation

4.6 Control

The human body has ~10^{16}, ten thousand million million cells, each performing its function like a colony of ants. They're controlled by a hierarchy of glands issuing hormones, some boosting cell division and growth, others causing cell death, apoptosis.

Copper and manganese

A gland below the brain, the hypothalamus releases nine kinds of Cu-containing hormones, targeting the nine parts of the anterior pituitary gland Fig 4.27:

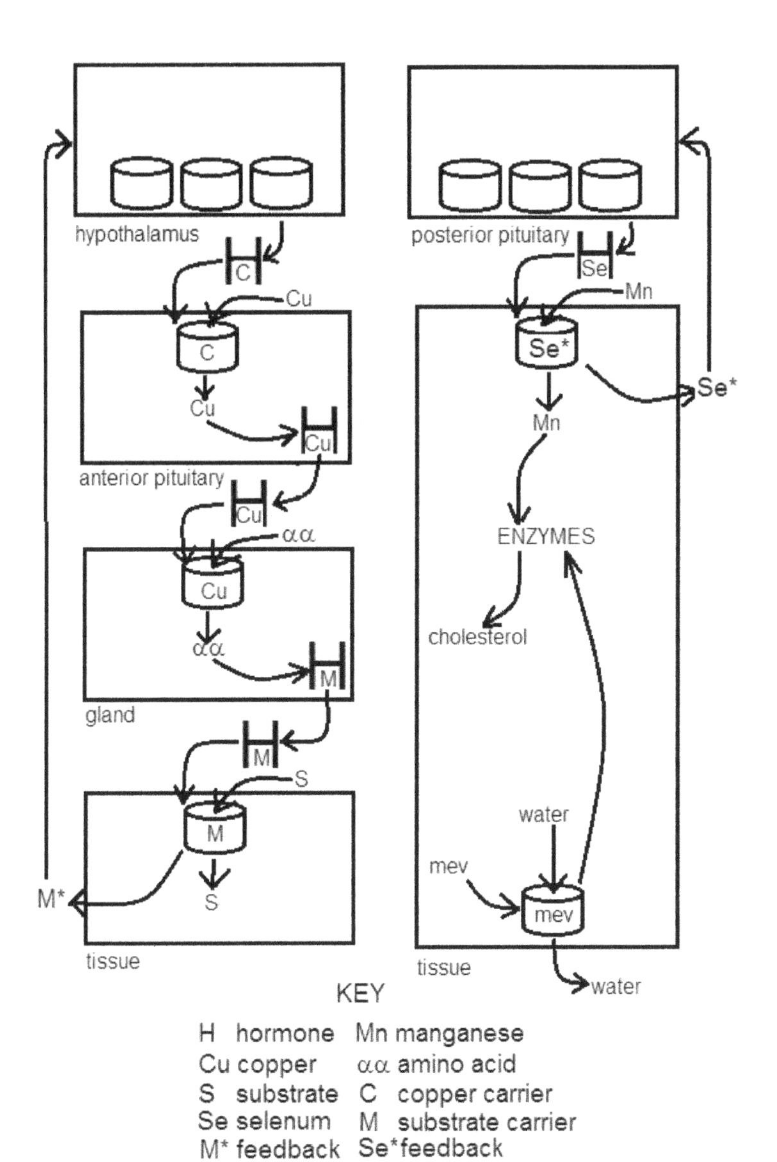

Fig 4.27 Cu/Mn control cell division/water transport

The anterior pituitary produces two families of hormones, one group packs copper, Cu^{++} ions into growth hormones which prime the pumps of endocrine glands to uptake amino acids and synthesize a third generation of hormones, priming the tDNA pumps of the nine tissue categories introduced in chapter 5, Fig 4.28:

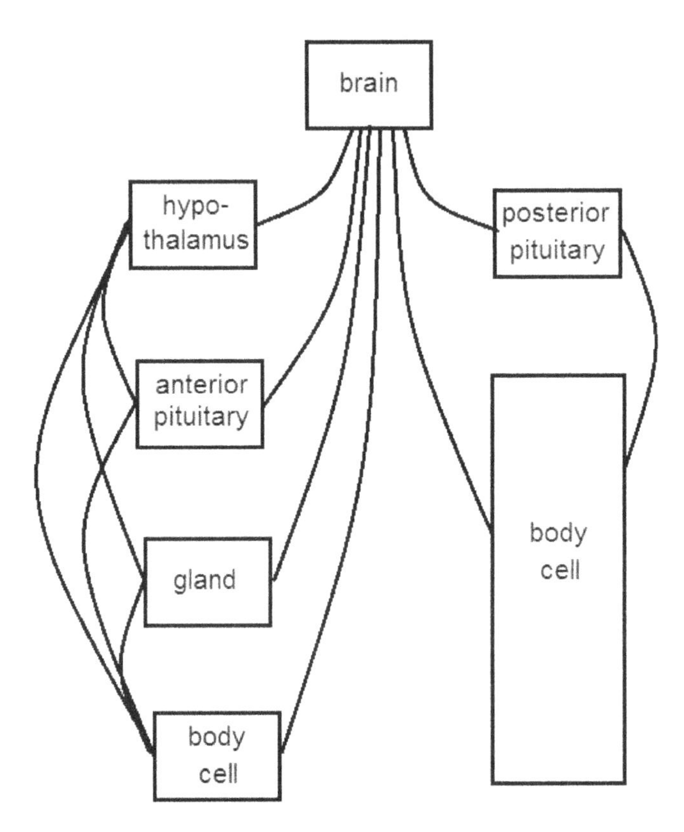

Fig 4.28 Feedback control of tissue differentiation

The other anterior pituitary hormone family carry manganese, Mn^{++} ions, priming chloride, Cl^- uptake, see section 5.3. This breaks down proteins and causes cell death, apoptosis, balancing the cell proliferation attributable to the Cu-carrying hormones.

Manganese also has a role in pumping water. The posterior pituitary packs selenium, Se into two hormones, vasopressin and oxytocin controlling blood pressure and water pumping, osmoregulation, see section 5.9. Vasopressin's targets are the colon, breasts, prostate and tear glands. Oxytocin focuses on reproductive organs.

At their targets, vitamin E, α-tocopherol, converts Se to selenite, $SeO_3^{=}$ which carries manganese, Mn^{++} into the cell, activating enzymes converting mevalonic acid into cholesterol.

Amplification

The three hormone types described above are issued by hypothalamus, pituitary and endocrine glands, each amplifying the next. One Cu atom leaving the hypothalamus causes a pituitary gland cell to take up amino acids, creating 100,000 hormone molecules. Each hormone molecule causes an endocrine gland to create 100,000 hormone molecules. These hormones activate cells, promoting transport of a million substrate molecules. This chain reaction amplifies the Cu atom's signal 10 million billion fold, by a factor of 10^{16}. Likewise, one Mn atom leads to 10^{16} recycling operations. In each case, one atom influences all the 10^{16} body cells.

4.7 Environment

Petroleum availability encouraged the development of internal combustion and diesel engines to power *horseless carriages*. Oil reserves will eventually be exhausted. Food supplies and the environment are threatened by converting food crops into alcohol for fuel and extracting gas from oil shale. Society can ill-afford the luxury of private transport if public transport uses fuel more economically.

Environmental variables

Preserving the natural environment is paramount when choosing a source of income, where to live, what to eat or how to enjoy exercise, entertainment, recreation and vacations. As the human population size increases, 'a cow and an acre' is no longer everyone's right. Education about birth control, healthy eating and the nine independent, homeostatic systems introduced in chapter 5 could lead to better ways to share the planet.

If wilderness is preserved for wildlife, land use for cities, agriculture, mining and oil wells is limited. Their pollution, pesticides, slag-heaps and oil spills must be controlled. Air, land and water transport must respect the environment and alternative energy sources be developed. As technologies advance, few variables are beyond our control. Soil fertility can be maintained, climate change averted and pollution prevented by adjusting our lifestyles.

We need care for our health by adopting balanced, varied diets, taking exercise, relaxing and taking advice from doctors, psychiatrists and nutritionists. Elected governments can assure our freedom of choice, basic rights and welfare, educate our children and protect us from fires, floods and epidemics. This book explores ways to balance human lives with other life forms.

4.8 Copying

Reproduction

Every form of life reproduces by copying nucleic acids, DNA base-pairing makes this straightforward, see 1.3, 7.4. Propagating the species is a primitive drive, whether by sexual intercourse or pollinating flowers. Accurate copying, giving the offspring high quality copies of their parents' DNA, ensures their fertility, Figs 4.29 and 4.30.

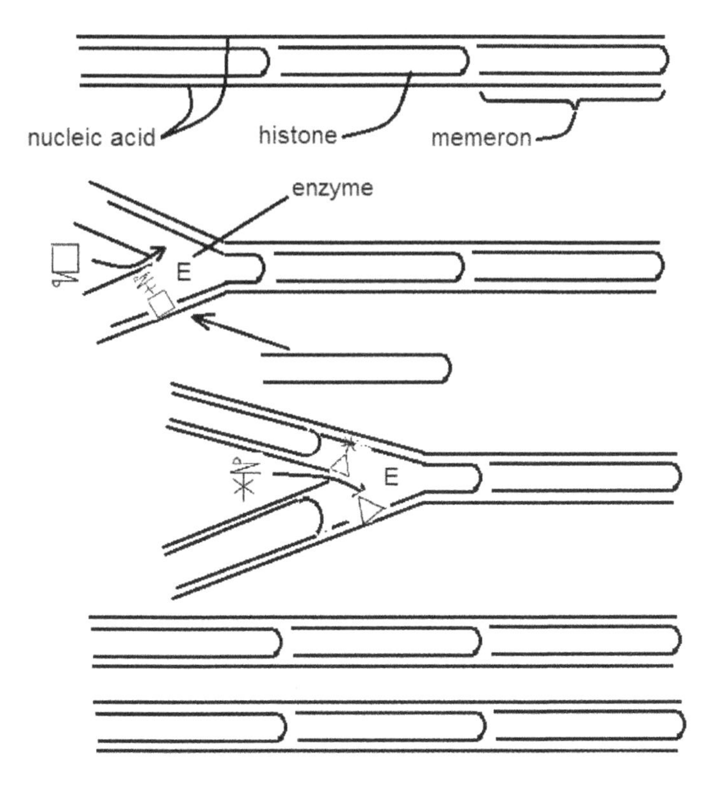

Fig 4.29 DNA replication

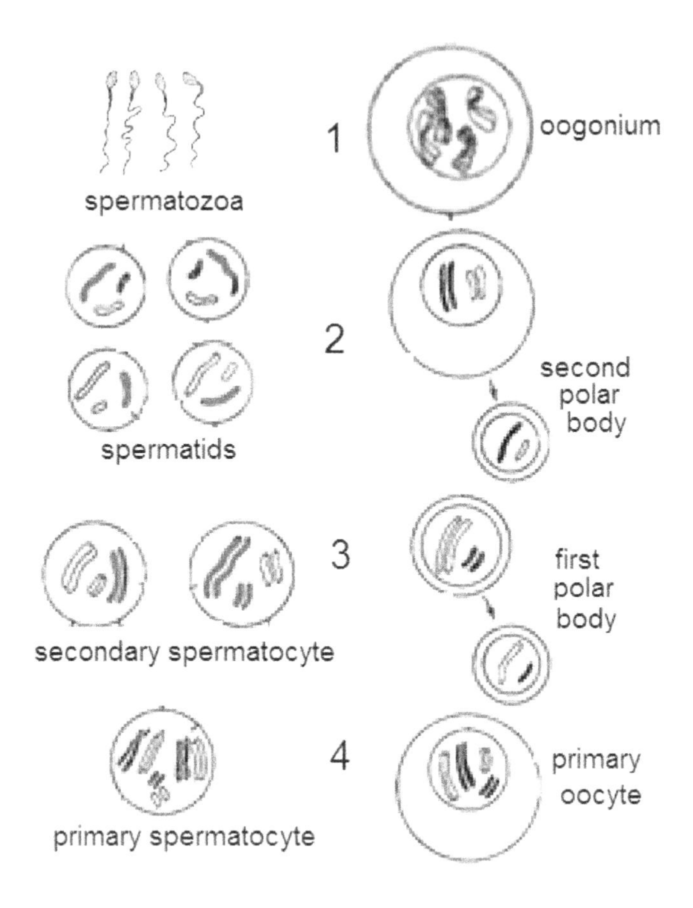

Fig 4.30 Egg and sperm formation

Mimicry

To construct an electronic entertainment system, logicians, electrical engineers and craftsmen cooperate. A cabinetmaker assembling the casing need know nothing about physics. The mathematicians designing the circuit could be oblivious of aesthetics. The soldering iron technician or robot replacing him may be ignorant of binary logic.

Patent office clerks comparing the product's features with previous examples protect intellectual property rights against all forms of mimicry. Users complain if devices don't obey familiar rules, if pressing a recognized button doesn't yield the expected outcome. They're deterred by complicated instructions. Conflict of interest, cosmetic changes and obsolescence are outlawed.

If a pharmacist doesn't understand biochemistry, the drugs he creates may have side effects, adding to patients' troubles. Failure to follow nature's rules, established by evolution, also needs outlawing. Progress is made by finding ways to make medicines conforming to natural requirements, recycled without polluting the environment and not encouraging mutant pathogens to arise, see sections 5.1–5.9.

4.9 Evolution

Genetic stock

Chapter 6 explores evolution in detail, describing how diverse animals, plants and micro-organisms arose from a common ancestor. The ways predators are outwitted, infections resisted and resources efficiently used are fascinating. *E.g.* anglerfish create light in the dark ocean depths, bats find their prey by echolocation and pitcher plants drown insects to obtain nitrogenous nutrients. All these tricks are encoded using the same tRNA and tDNA alphabet; they could be transferred to other species.

Genetic engineering

Changing tDNAs or tRNAs could have far-reaching consequences. Modifying mRNAs encoding proteins is safer. To make significant improvements, several mutations may be necessary, differing between species. *E.g.* a mutation improving a shark's gills might enhance a sparrow's lung capacity and zebra fish genetically modified using jellyfish genes glow in the dark. Genetic engineers' powers are limited, they're unlikely to teach buttercups to ride bicycles or enable sunbathers to photosynthesize.

Genetic balance

Animals and plants diverged long ago, creating an animal that could photosynthesize would be difficult. Ecosystems' components depend on each other for survival. Animals rely on plants converting carbon dioxide into oxygen, see 5.4. Plants rely on animals to distribute their seeds. Failure to respect ecological balance disrupts the natural environment. *E.g.* eradicating bees could render some flowering plants extinct. Changes need take every species into account.

5 In sickness and in health

Preface

Our body maintenance contracts are non-transferrable, requiring defects to be corrected. If we know ourselves and seek advice, we recognize when something is amiss, acting on the diagnosis. This chapter describes nine parallel systems controlling the body. Most disorders relate to just one system, its endocrine glands or the trace elements involved in its function. Many health problems arise from unbalanced diets. Identifying the affected system facilitates diagnosis. I offer insights, not instructions for home medicine.

5.1 Motility

Calcium, magnesium and sulphur

The majority of cell functions are driven by magnesium, Mg^{++}, catalyzing hydrolysis of ATP, Fig 5.1. Calcium, Ca^{++} inhibits it, sulphite, $SO_3^{=}$ exchanges Ca^{++} and Mg^{++} using a tDNA pump, Fig 5.2.

$$Ca^{++} + SO_3^{=}.Mg^{++}.\ SO_3^{=} \leftrightarrow Mg^{++} + SO_3^{=}.Ca^{++}.\ SO_3^{=}$$

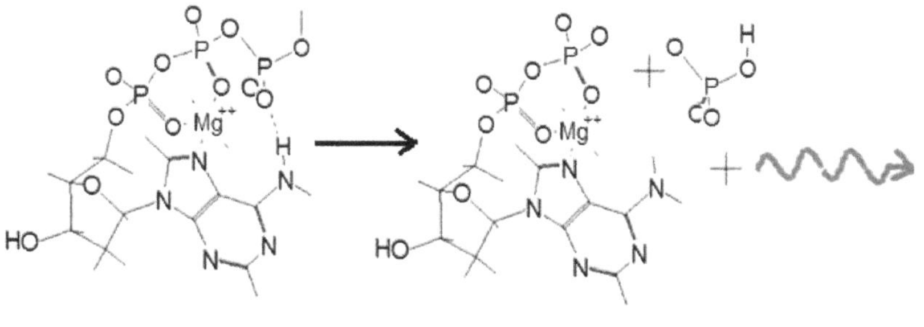

Fig 5.1 ATP hydrolysis

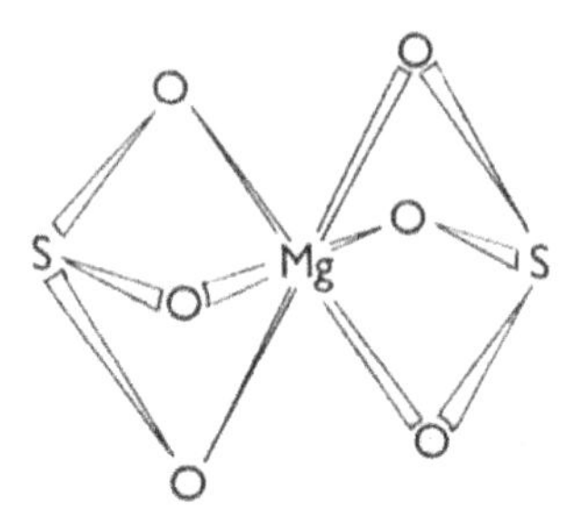

Fig 5.2 Sulphite binds magnesium

Sulphur and magnesium deficiency are rare. Attention to diet can correct the cramps, arthritis, nervous agitation and movement difficulties arising from deficient Mg. It's a component of chlorophyll, enabling plants to photosynthesise. ATP hydrolysis powers animal muscles, bacterial flagella and sperm tails.

Cell charge

Most active transport complexes are singly-charged. Mg^{++} and Ca^{++} complexes with $SO_3^{=}$ are doubly-charged, making exchange of Mg^{++} and Ca^{++} sensitive to membrane potential. Hydrolysing acetyl choline to choline for uptake enhances this. Cholinesterase inhibitors, Fig 5.3 block it.

If S, Mg and Ca are sufficient, acupuncture or a whiff of ammonia may solve motility problems arising from cell charge. Take drugs only if a doctor advises you to do so.

Cholinesterase releases the giant cation choline from acetylcholine. Choline uptake changes cell membrane potential. Cholinesterase inhibitors prevent Mg^{++}/Ca^{++} exchange and energy release from ATP, they've been used for chemical warfare.

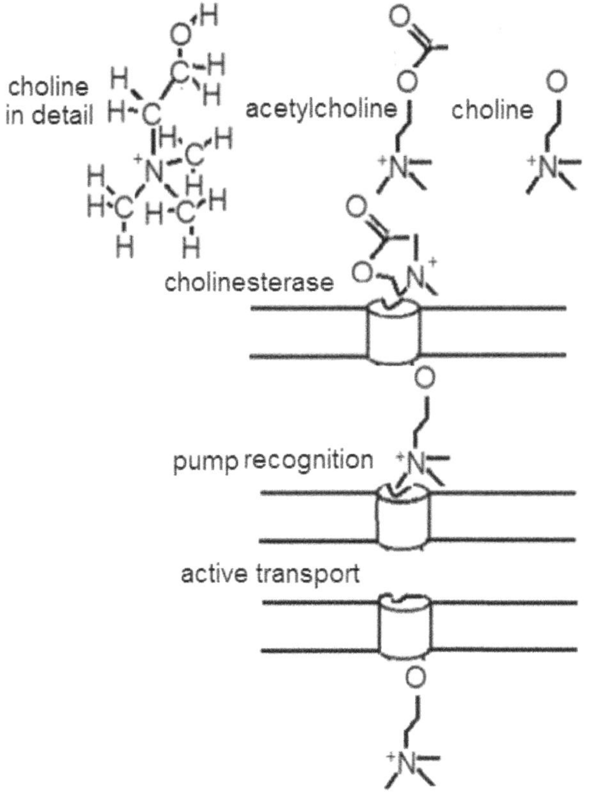

choline in detail

acetylcholine

choline

cholinesterase

pump recognition

active transport

Fig 5.3 Choline esterase releases choline

5.2 Nerve transmission

Sodium and potassium

Fig 5.4 shows sodium ions, Na^+ creating clouds of sodium hydrate, $Na^+.28H_2O$, it converts cell sap to jelly. The increased cell viscosity slows all the cell's chemical reactions. Gold, Au^+ makes a sol which can substitute for Na^+, some made by Michael Faraday still exists.

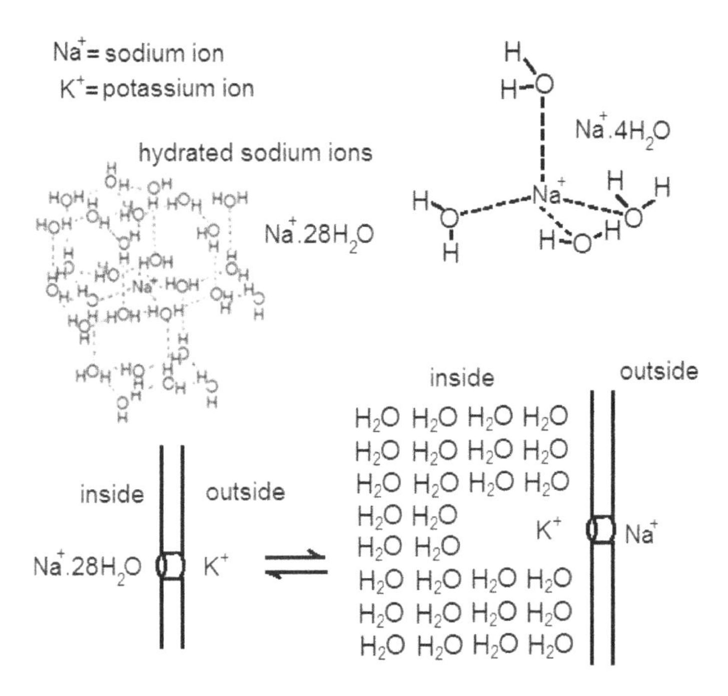

Fig 5.4 Sodium binds more water than potassium

Complexes between adrenal hormones: adrenaline, noradrenaline and dopamine form four- and six-member rings around K^+ and Na^+ ions, exchanging two K^+ for three Na^+. They transmit nerve signals by changing cell charge, Fig 5.5:

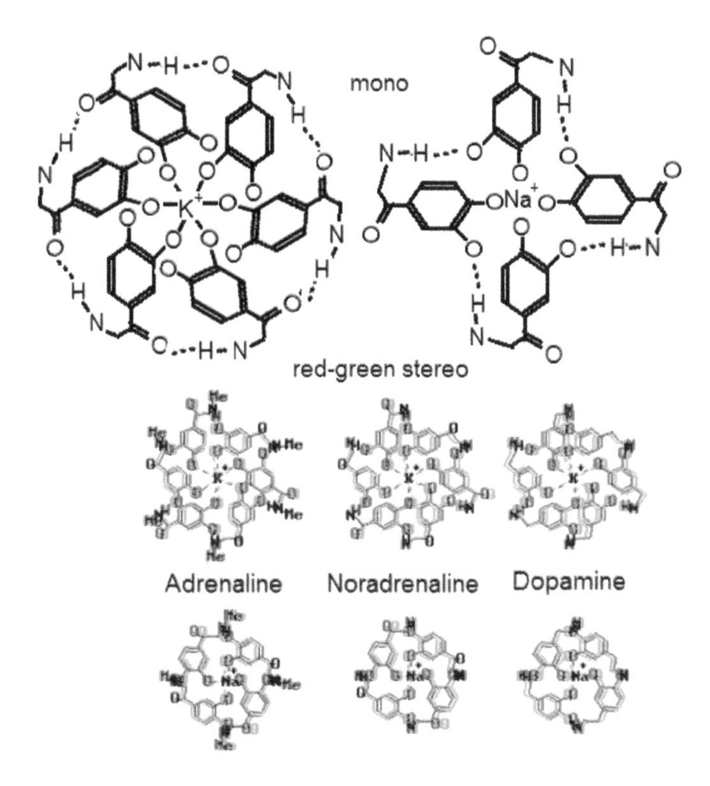

Fig 5.5 Catecholamines swop 2 K^+ for 3 Na^+

Attention is increased when we're frightened, explaining the *fight or flight* reaction associated with adrenaline. Complexes formed when morphine or codeine substitute for adrenaline are larger, blocking the tDNA and suppressing pain, Fig 5.6. When more tDNAs are enrolled to compensate, pain sensitivity increases, accounting for drug addiction.

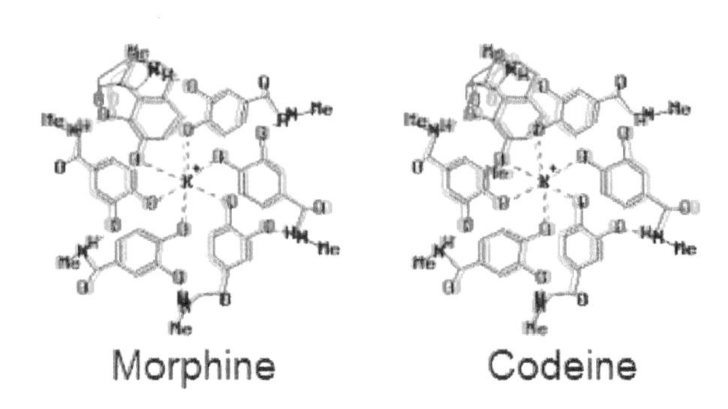

Fig 5.6 Morphine and codeine mimic catecholamines

Nerve transmission

α- and β-adrenergic receptors are pumps exchanging Na^+ with K^+, many drugs target them. They engage bottom gear during sustained mental or physical work, sapping strength, and slowing mental and physical activity. Meditation, relaxation, soft lighting, calming music, favourite foods, pleasant scents and massage can all restore vitality. Society prefers self-reliant, lion-hearted, iron-willed believers in golden truth to slaves to routine.

Reproduction

'Royal jelly' creates queen bees, low Na^+ levels at cell division facilitate chromosome migration. Reproductive organs stiffen due to restricted blood flow for intercourse. Blood's high viscosity resembles that of the seawater life once inhabited.

Pathology

$tDNA_{Na/K}$ pump failure causes muscular dystrophy; defects of brain areas using dopamine as carrier for exchanging Na^+ with K^+ cause Parkinson's disease. Promiscuous sexual activity invites venereal diseases, VD.

5.3 Excretion

Manganese and chlorine

Manganese, Mn^{++}, a rarely deficient trace element, controls sodium chloride, NaCl transport, maintaining osmotic pressure and cell size. Chloride, Cl^- mediates cell death, apoptosis, Fig 5.7:

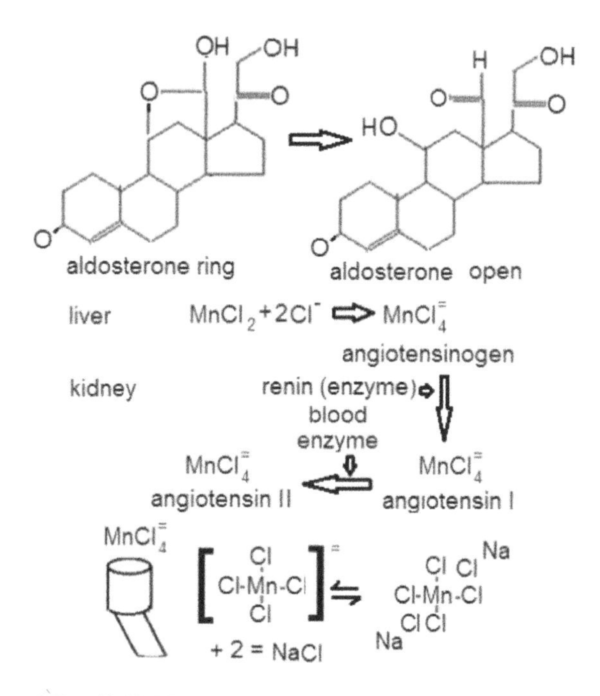

Fig 5.7 Manganese transports salt

Exchanging Cl^- with bicarbonate, HCO^{3-} controls red blood cell acidity, pH. Salt needs regular replenishment in hot weather to compensate for that lost in sweat. Hydrochloric acid, HCl digests food in the stomach. Chlorine, Cl_2 is used to disinfect water, making it safe to drink and countering poor hygiene causing infection from excreta.

Skin

The skin, our first line of defence against sharp objects, heat and cold is constantly replaced. Its fragments contribute substantially to household dust. Cuts, abrasions and blisters expose us to bacterial infections and the gut lining is susceptible to ulcers. Antibiotics can help if cleaning doesn't heal them. Care and hygiene compensate for excretory system deficiencies. New foods keep stomach ulcers at bay.

Heat, cold

Hair regulates our temperature and protects us from damaging ultraviolet, UV light. Without the protection skin and hair afford, we'd evaporate in the wind or bake in the Sun; cuts and abrasions could hurt us. Excretions compensate for these dangers; sweat keeps us cool and tears clean our eyes.

5.4 Respiration

Oxygen, O_2 molecules are normally hydrated. $O_2.H_2O$ forms an isosceles triangle of oxygen atoms, binding purple, metallic iodine, I^+ for cellular O_2 import. The regime may have evolved from seaweed concentrating I to compensate for tidal O_2 fluctuations. The thyroid gland packages iodine into thyroxin, Fig 5.8.

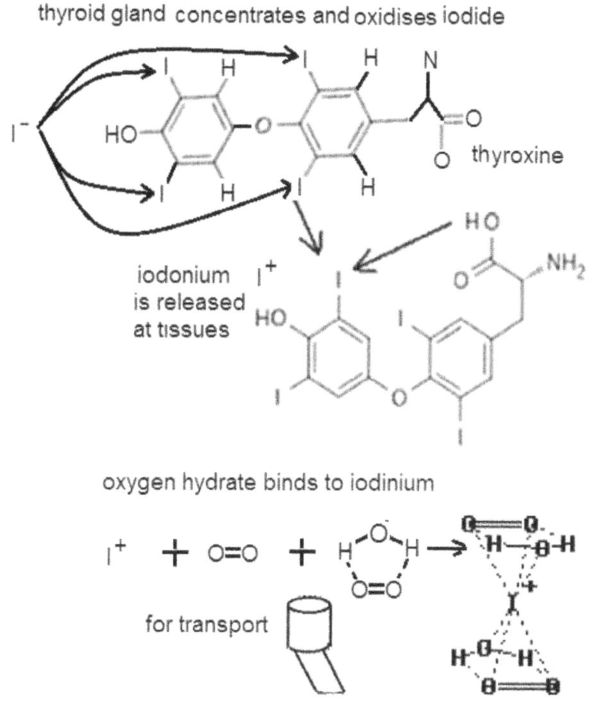

thyroid gland concentrates and oxidises iodide

thyroxine

iodonium is released at tissues

oxygen hydrate binds to iodinium

for transport

Fig 5.8 Iodine transports oxygen

Mitochondria are cell compartments just the right size to retain 4μ radiation, as discussed in Part 3. Mitochondria release the energy stored in carbohydrates and fat via a series of oxidation reactions, the 'citric acid cycle', charging ATP. Energy passes between a series of porphyrin rings (cytochromes), releasing 4μ quanta, ~→ at each step, Fig 5.9.

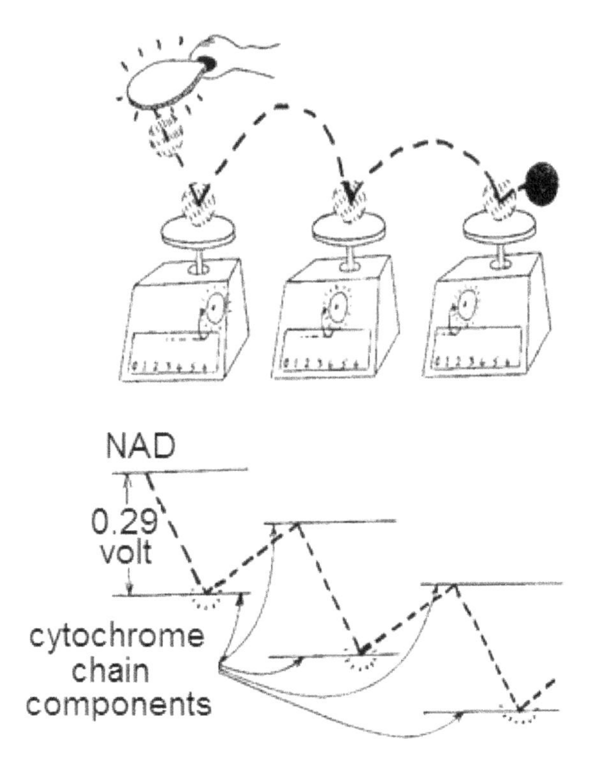

Fig 5.9 Cytochrome chain splits light into 4μ infrared

Porphyrin complexes are familiar: green chlorophyll with magnesium in plants and red haem with iron in red blood cells, others are involved in respiration, Fig 5.10.

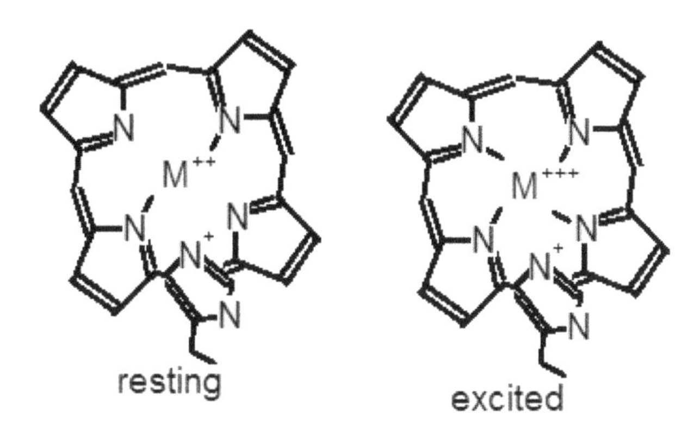

Fig 5.10 Haemoglobin, chlorophyll and cytochromes contain porphyrin rings

The nicotine constituent of nicotinamide adenine dinucleotide, NADH shuttles a proton, H^+ across the mitochondrial membrane to another nicotine molecule, forming NAD^+, Fig 5.11.

$$NADPH + NAD^+ \leftrightarrow NADH + NAD^+ + P_i$$

The enzyme cytochrome oxidase catalyzes the addition of oxygen radicals to hydrogen, converting oxygen, O_2 to water, H_2O.

$$2H + NAD^+.O \leftrightarrow NAD^+ + H_2O$$

$$\underline{H^{\pm} + NADH \leftrightarrow NAD^{\pm} + 2H}$$

$$H^+ + NAD^+.O + NADH \leftrightarrow 2NAD^+ + H_2O$$

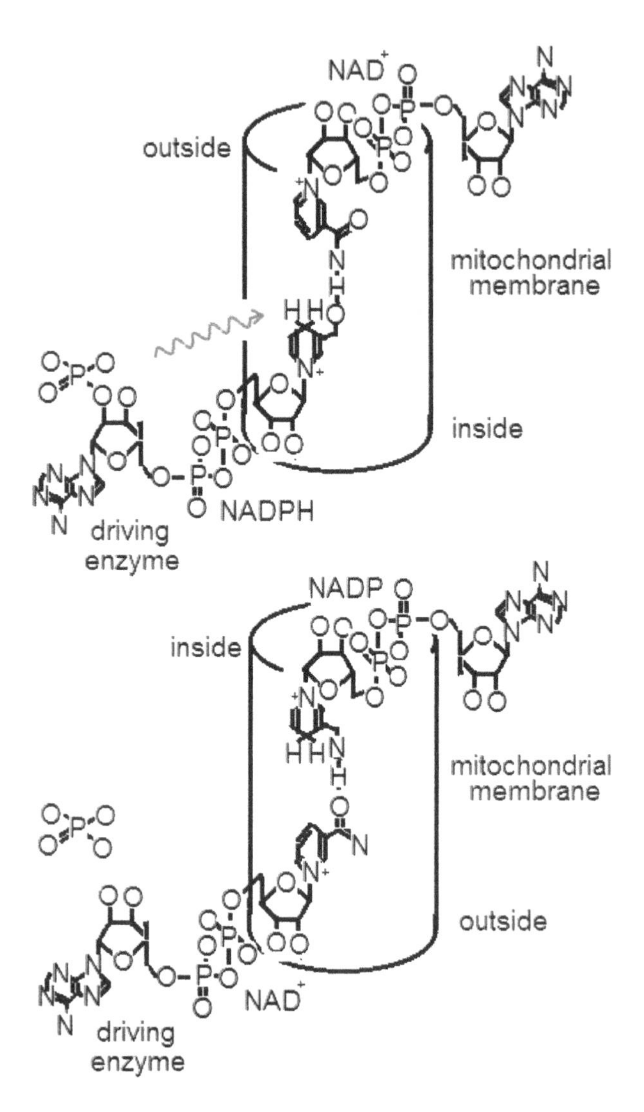

Fig 5.11 Nicotinamide of NAD transports protons

NADH fixes nitrogen and accepts oxygen and carbon monoxide; cyanide poisons it, Fig 5.12.

$$NADH + O_2 + H^+ \leftrightarrow NAD^+.O + H_2O$$

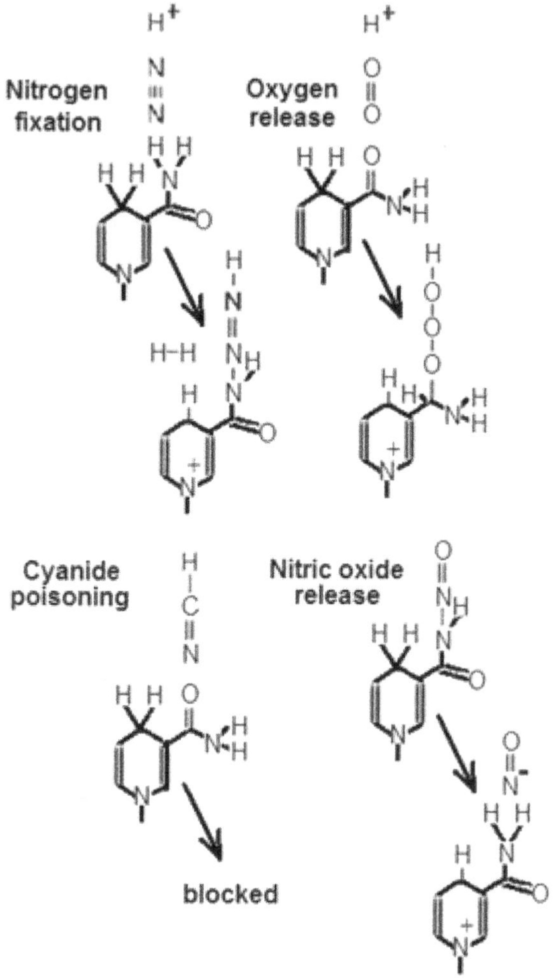

Fig 5.12 Nicotinamide of NADP drives N_2 fixation, O_2 and NO release subject to cyanide poisoning

NADH fixes nitrogen, Fig 5.13.

$$NADH + N_2 + H^+ \leftrightarrow NAD^+.N{=}N + H_2$$

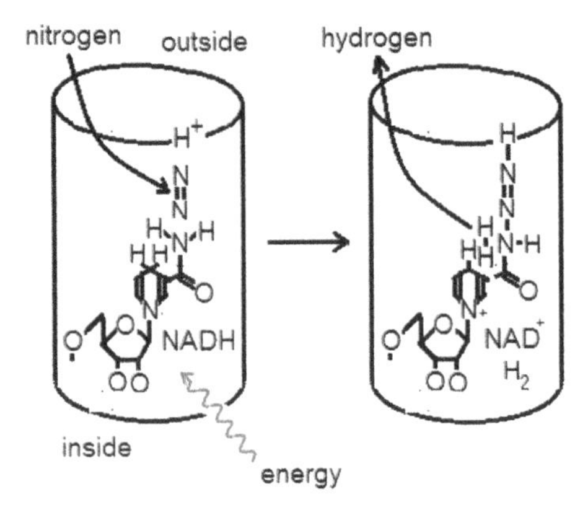

Fig 5.13 Nitrogen fixation

CO_2 accumulation triggers breathing, loading red blood cell haemoglobin with O_2, it acts like a sponge to distribute it. Blood cells are replenished in the bone marrow. Chlorophyll in plants absorbs sunlight energy, making sugar from CO_2 and H_2O.

Sugar acts as a battery, it's oxidised to release energy as needed. Animals store the sugar polymer glycogen in their livers, plants store sugar as starch.

Pathology

The highs and lows of bipolar disorder, *aka* manic depression, arise when an altered, mutated tDNA feeds too much or too little O_2 to brain cells. Lithium ions, Li^+, having the same shape and size as I^+, mimic it. Lithium got its name from the Latin *lithos*, a stone – it stabilizes the condition by maintaining steady O_2 levels.

Genetic engineers may be able to eliminate hereditary illnesses arising from mutant tDNAs. Greater social acceptance would make mental disorders easier to tolerate. Nervous breakdowns can be prevented by deep breathing, relaxation and meditation counter over-exertion. City parks and vacations afford opportunities for exercise, many organisations offer retreats for meditation.

Smoking tobacco augments nicotine, associated tars cause lung cancer. Tonics contain vitamin B_{12}, a porphyrin-cobalt complex. Both cyanide, CN^- and carbon monoxide, CO poison the oxygen transport tDNA. An organised life can eschew drug and tobacco use.

5.5 Growth

The most obvious answer to my question *What's natural?* is growth mediated by copper, Cu^{++} transporting amino acids, ααs for building proteins, *c.f.* Cu^{++} interacting with ααs in the Biuret test for protein.

The several components of the hypothalamus, a gland at the base of the brain, produce protein hormones, causing corresponding parts of the anterior pituitary gland to uptake Cu^{++} and synthesize hormones, see Fig 4.27, they stimulate endocrine glands to secrete more hormones. Sources of ααs include brown bread, nuts, lean meat and milk, Fig 5.14.

High protein meals are digested by bile secretions releasing ααs in the gut, they're best utilized during sleep, optimal time for protein synthesis. Gut bacteria aid digestion and synthesize essential vitamins. The kidneys excrete surplus ααs as urea in urine. The liver equilibrates different types of αα, supplying the brain with a balanced diet. The brain-gland-liver triangle and regular αα intake regulate our state of mind, if in doubt sleep on it!

Cu contraceptives compete with Zn by inhibiting sperm's access to sugar, see section 5.7. When contraceptives and chastity belts fail, unwanted pregnancies can be aborted. We make sacrifices for love, nurture placenta and foetus and provide breast milk to infants.

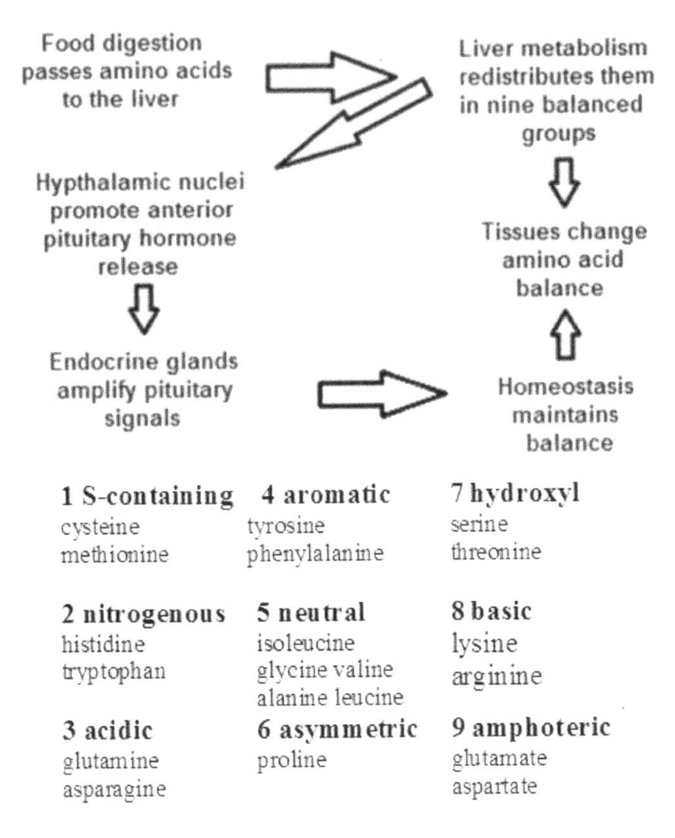

Fig 5.14 Copper controls tissue differentiation and protein synthesis

Attitudes and feelings, touch, instinct, sleep and reliance on others support growth. Totalitarian regimes' policies on population control can be extreme; Winston Churchill said *'With opportunity comes responsibility'*. Growth thrives on small units, whether of proteins or families. *'No man is an island'*, seek help from teachers and physicians, be vigilant and rely on others.

Pathology

Smoking, drinking and drug abuse provide unusual proportions of nitrogen compounds, upsetting hormone balance. Cigarette smoke fertilizes the lung surface, like nitrogenous fertilizers concentrated in lakes and seas filling them with weeds. Nicotine, alcohol and morphine cause cancers and tumours. Excess alcohol over-stimulates the liver, causing a cheesy growth, cirrhosis.

Mosquito nets, water filters and food purification prevent parasite infections by worms and malarial mosquitoes. Many religions offer advice; cures can be painful.

Iodine and copper accumulating in the eyes can cause exopthalmos, pop-eye and Wilson's disease, *c.f.* zinc causing glaucoma, see 5.7. Growth defects need careful monitoring, they develop slowly and are hard to detect, causing general feelings of ill health. Preventing cancers, tumours and malformations is cheaper than surgery. Antibiotics could be designed to target particular tDNAs.

Supplementing the blood's Cu reserve can relieve arthritis. Troubles with this system need be met halfway; common sense and the instinct to survive dictate the best approach to unwanted pregnancies or unsightly growths.

5.6 Rigidity

Skeletons

Animals have rigid bone and tooth skeletons, elephants have ivory tusks, sea creatures' and birds' eggs are protected by carbonaceous shells and diatoms have silica skeletons. Skeletal materials serve as levers, props, cages, boxes, knives, grinding tools, shields and swords. Tiny bones transmit sound in our ears.

Exposure to ultraviolet, UV sunlight energizes vitamin D, its absorption spectrum accords with Si-F binding energy. It enables the parathyroid glands in the neck to load parathyroid hormone with silicon hexafluoride, $SiF_6^=$, Fig 15.

$$SiO_2 + 6HF \leftrightarrow SiF_6^{=} + 2H^+ + 2H_2O$$

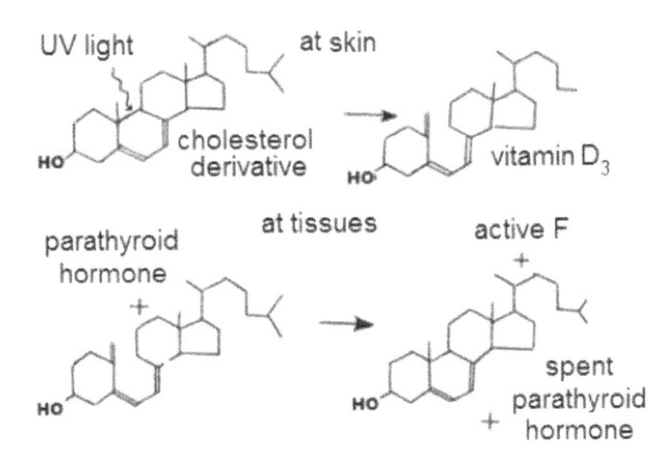

Fig 5.15 UV light energizes vitamin D, parathyroid hormone controls $SiF_6^{=}$ synthesis

Octahedral $SiF_6^{=}$ binds Ca^{++}, **a** and **b** forming a tetrahedron, **c** and **d** before collecting $PO_4^{\equiv}$, **e**. Adding either a hydroxyl group, OH^- or fluoride, F^- completes the complex, Fig 5.16.

$$HOCa^+.3PO_4^{\equiv}.4Ca^{++}.\ SiF_6^{=}.4Ca^{++}.3PO_4^{\equiv}.CaOH^+$$

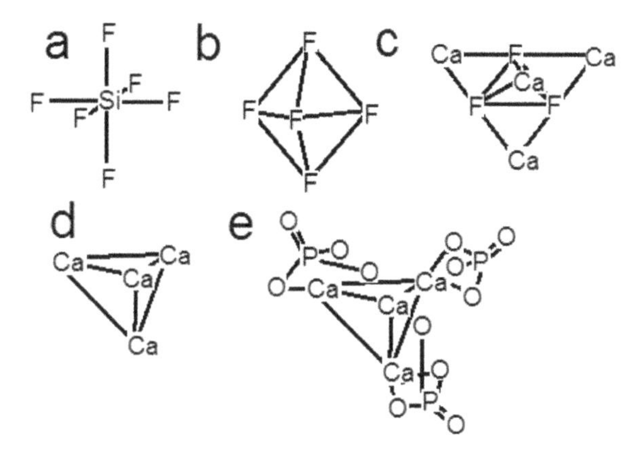

Fig 5.16 $SiF_6^=$ binds calcium phosphate for bone and tooth maintenance

$SiF_6^=$ carries apatite, calcium phosphate for building bones; diatoms use apatite as carrier to build their silica, SiO_2 skeletons, preserved as diatomaceous earth. These processes and marine shell growth are pH-sensitive – at risk from climate change acidifying the oceans and atmosphere.

Attitudes

The body responds to repetitive exercises such as piano playing or gymnastics stressing bones by rebuilding and strengthening the skeleton. Blows to the head endanger the brain; skiing accidents can break a leg. Wheelchairs, false limbs and teeth needn't change your outlook on life. Don't carry a chip on your shoulder for failing an examination. Adopt daily routines, form steady relationships, abstain from over-indulgence and take exercise to build your appetite.

Pathology

Vitamin D deficiency in childhood causing rickets, leading to such skeletal deformities as knock-knees and bow legs, is easily prevented by cod-liver-oil. Fluoride toothpaste and F^- in drinking water strengthens tooth enamel, converting apatite to hard fluorapatite and preventing tooth decay. Excess F^- causes mottled tooth. Without grit, chickens lay soft shelled eggs. Cups of tea supply adequate F^-.

If the olfactory nerve endings at the back of the nose are acidified at menopause, in kidney failure, by tobacco smoke, sulphur or nitrogen dioxide, SO_x or NO_x air pollution, $SiF_6^=$ is inappropriately synthesized and carried into the brain. Its breakdown there releases F^-, killing brain cells and leading to Alzheimer's disease, fluorinated anaesthetics temporarily relieve it. They promote F^- excretion as aluminium hexafluoride, AlF_6^{3-}, simultaneously clearing F^- from the brain.

Make each day special for yourself or your loved ones, take opportunities, not risks, keep an open mind and exercise your body, take *From this day forth* as your motto. Look back in anger and forward with longing; society is the best medicine for sustaining progress. Whatever your debilities, accept some responsibilities, stand up for yourself and participate in community life.

5.7 Assimilation

Appetites for sweet or bitter tasting food and milk or honey are normal, the calorie content of sugar and fat supplies energy, whether we feel happy or blue. Hunger drives the world to work, insobriety invites dismissal. Obesity demands self-discipline.

When food is detected in the gut, the pancreas packs zinc, Zn^{++} in insulin, distributing it throughout the body. Vitamin C, ascorbic acid from fresh fruit, is metabolised to form $\beta\text{-}_L$gulonate, delivering Zn^{++} to regions insulin can't reach. Another hormone, glucagon, returns it to the pancreas. Zn^{++} transports $\beta\text{-}_D$glucose, the commonest kind of sugar, Fig 5.17.

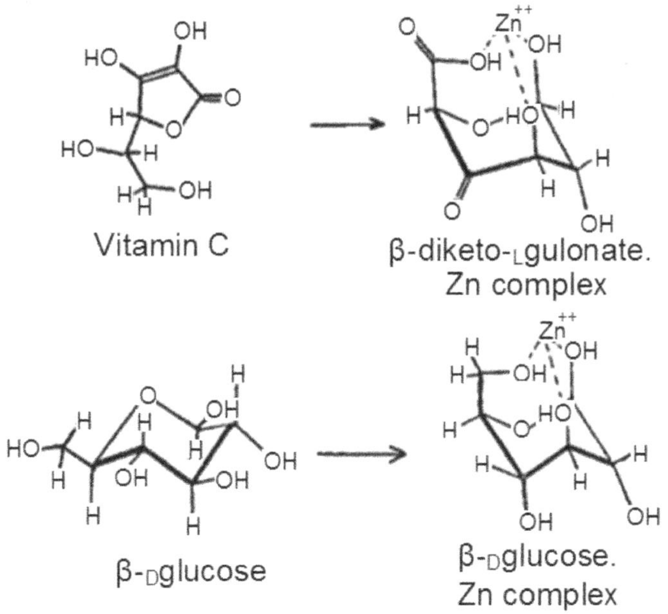

Vitamin C → β-diketo-$_L$gulonate. Zn complex

β-$_D$glucose → β-$_D$glucose. Zn complex

Fig 5.17 β-$_D$glucose binds Zn and vitamin C derivative β-diketo-$_L$gulonate delivers it to tissues inaccessible to insulin

The gut releases glucagon when food is anticipated, disabling all zinc-glucose pumps and rendering the liver receptive to glucose, storing it as glycogen and slowly releasing it to maintain a constant blood sugar level. Stresses and strains promote insulin release from the pancreas, recharging the pumps supplying cells with sugar. Alcohol disrupts the system by moving Zn^{++} to the liver for oxidation by the Zn-dependent enzyme alcohol dehydrogenase. Barbiturates have the same effect, the combination of alcohol and barbiturates can be lethal.

The Victorians were poisoned by mendaciously administering beryllium, Be^{++} and the Romans by using lead, Pb^{++} as a sweetener, both mimic Zn^{++}. Zn is present in aphrodisiacs, promotes appetite and determines satiety, whether we feel full, inhibiting over-eating.

Fat builds membranes, its metabolism is pump-independent; acetyl coenzyme A controls fat metabolism. Fat deposits complement liver glycogen. Connective tissue proteins contain the amino acid hydroxyproline, Hyp. Hyp is incorporated into the proteins by Zn^{++} instead of Cu^{++}, skin, membrane and ligament maintenance depend on Zn.

Well nourished individuals have better figures, enhanced by fat, skin and hair. The skinny or obese shouldn't be mocked. When judging appetite, allow for emotions; good looks depend on what we eat and make of life.

Pathology

Zinc could be used to manage diabetes, prevent scurvy, colds, flu and alcoholism. Diabetics get little sympathy for being overweight, although the problem lies beyond their control. Organisations like weight-watchers, alcoholics and gamblers anonymous help sufferers by sharing problems and enhancing their self-confidence; a problem shared is a problem halved.

We need recognize our strengths and weaknesses and avoid running into debt, attempting to lift excessive loads and over-eating. Zn and vitamin C supplements counter one of our commonest troubles, catching a cold or influenza, by preventing $tDNA_{zinc\text{-}glucose}$ pumps taking up rhinoviruses. These supplements can also augment hormone injections and special diets in diabetes control and counter obesity and alcoholism. Encourage the gourmet in the sufferer to compensate for dull food and dispel feelings of abnormality.

Zn or vitamin C deficiency causes the skin lesions and infertility of scurvy; sperm can't reach their targets without Zn for sugar uptake. Afford friendship to the lonely, dissuade solitary drinking, overeating and overspending offer a shoulder for others to lean on and confidential advice. Don't lose control, depression can be infectious!

5.8 Energy

Silver iodide, AgI plates were once used to make photographic images; the nervous system is the canvas for our imagination. Silver, Ag^+ controls high-energy compounds. The pineal gland or third eye distributes it bound in 6-member serotonin rings, Fig 5.18 **a** resembling those adrenaline forms around potassium, see section 5.2. Serotonin, **c** is related to N-acetyl derivative, melatonin, **b** which regulates sleep, necessary for allowing the body to restore itself daily.

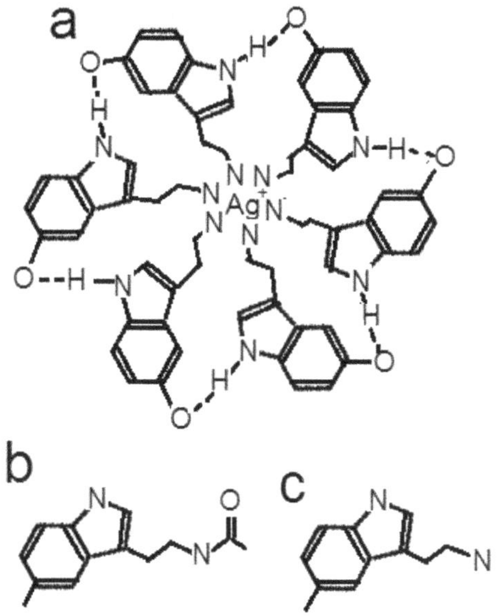

Fig 5.18 Serotonin carries Ag^+ from pineal gland

Vitamin A has two isomeric forms, retinol and retinal, Fig 5.19:

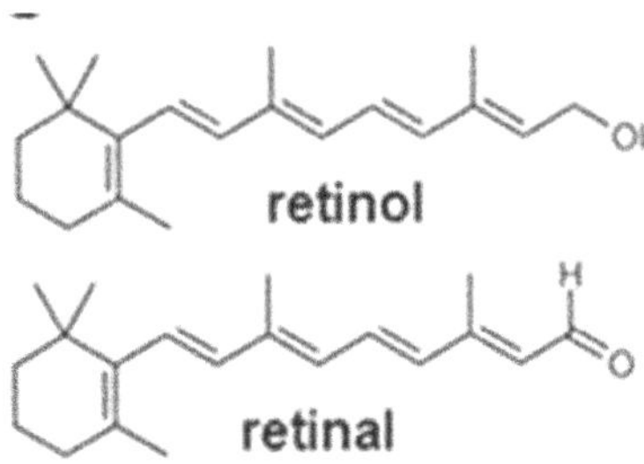

Fig 5.19 Isomers of vitamin A

The latter has alternate single and double bonds, carrying energy in the form of 'solitons' from silver porphyrin to convert phosphate, P_i to its dimer pyrophosphate, PP_i transported as a complex with arginine, Fig 5.20, mimicked by anti-cancer drugs:

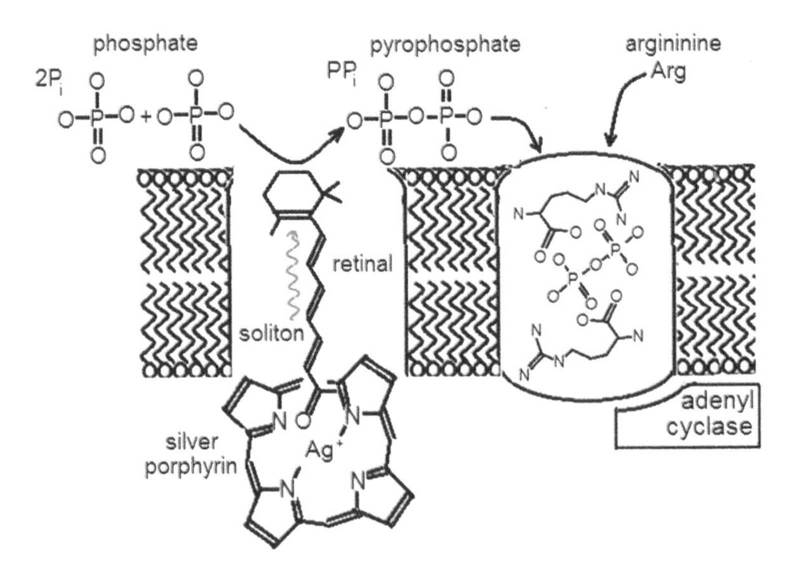

Fig 5.20 Energy from silver porphyrin assembles pyrophosphate for transport by arginine

You may remember the chemical test for Ag^+ is the white precipitate it forms with chloride, Cl^-, nonetheless biology manages to keep Cl^- at bay. ATP is too big to pass through a tDNA pore, creatine and Ag^+ transport phosphate, $PO_4^{\equiv}$, P_i, for energising the nucleic acid bases □, Δ, + and *, Fig 21. The creatine phosphate complex contains all the ingredients for synthesizing DNA. All energy dependent operations use high-energy compounds, normally ATP.

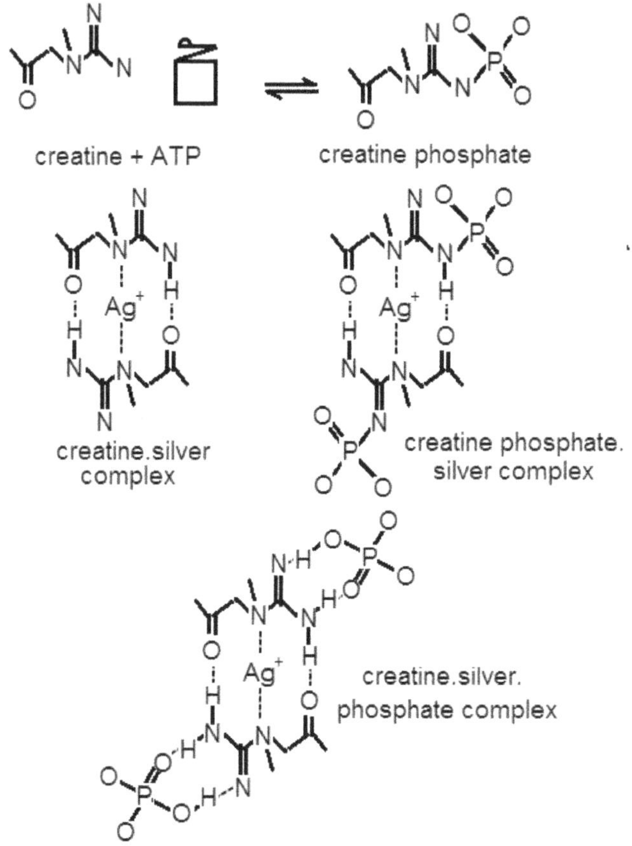

Fig 5.21 Silver transports DNA constituents

Getting angry or developing phobias and allergies wastes energy, beware of anger rising to fury and leading to hatred, crime and war. Depression can be relieved by drugs or by seeing the bright side of life. When free from worry and anxiety we have a better view of the future, seeing beyond the horizon by combining vision, experience and imagination.

The pineal is also known as the *third eye*, enacting our dreams and visions and balancing energy expenditure. The reliability of our ideas need be checked by thought and discussion before they're implemented. Beautiful surroundings stimulate the imagination and put our thoughts in perspective. Poisonous arsenate, $AsO_4^{\equiv}$ mimics $PO_4^{\equiv}$, chlorinated hydrocarbons like chloroform, $CHCl_3$ and the pesticide dichloro-**d**iphenyl-**t**richloroethane, DDT disrupt the system. Hypnosis counters allergies and phobias, just as anaesthetics enable surgeons to operate. Patients awake from each with more purpose, direction and order in their lives, *silver threads amongst the gold.*

5.9 Water regulation

Unlike fish immersed in salt water and plants using rigid cell walls, land animals lead salt-free lives. Goats don't get heart attacks leaping about on cliffs, fresh fruit and vegetables, *five a day* prevents them. Plants expend most of the energy received from sunlight to construct cellulose cell walls retaining water.

Water-pumping maintains steady blood pressure, combating osmosis, the tendency of salt to attract water, keeping the water pressure within cells constant. Mevalonate, named after the herb Valerian, *aka* All Heal carries water by reversibly forming a *lactone* ring, Fig 5.22.

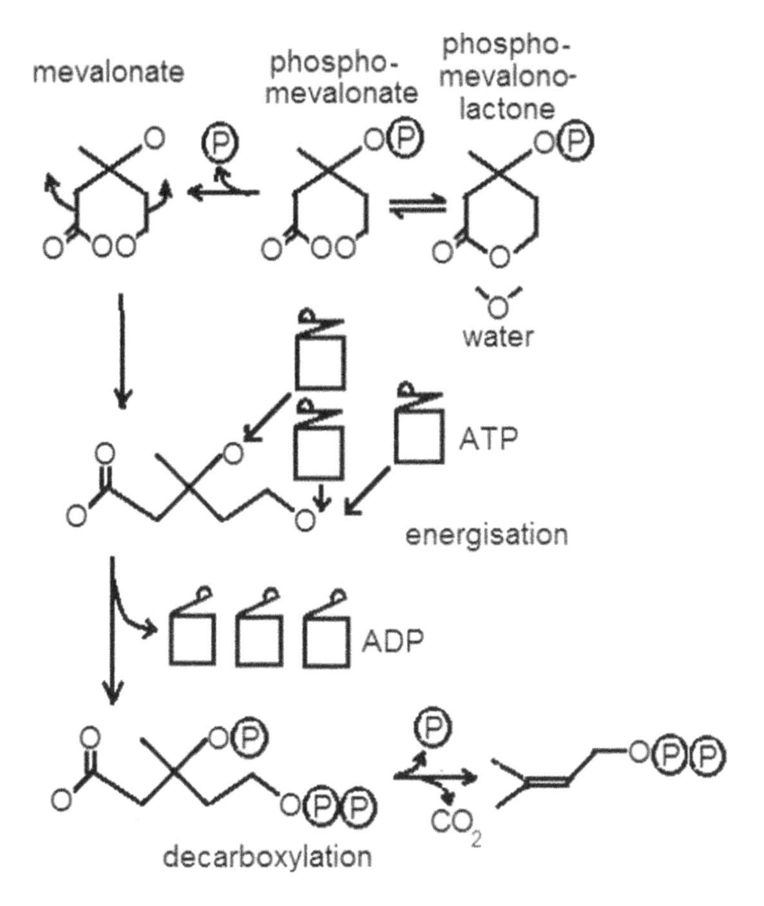

Fig 5.22 Mevalonate transports water, Mn catalyzes its conversion to cholesterol

The lactone behaves like a crab's claw holding and dropping it when clasping its pincers. Mevalonate is the residue of saturated fat breakdown; manganese, Mn^{++} catalyzes its polymerization, forming cholesterol, Fig 5.23.

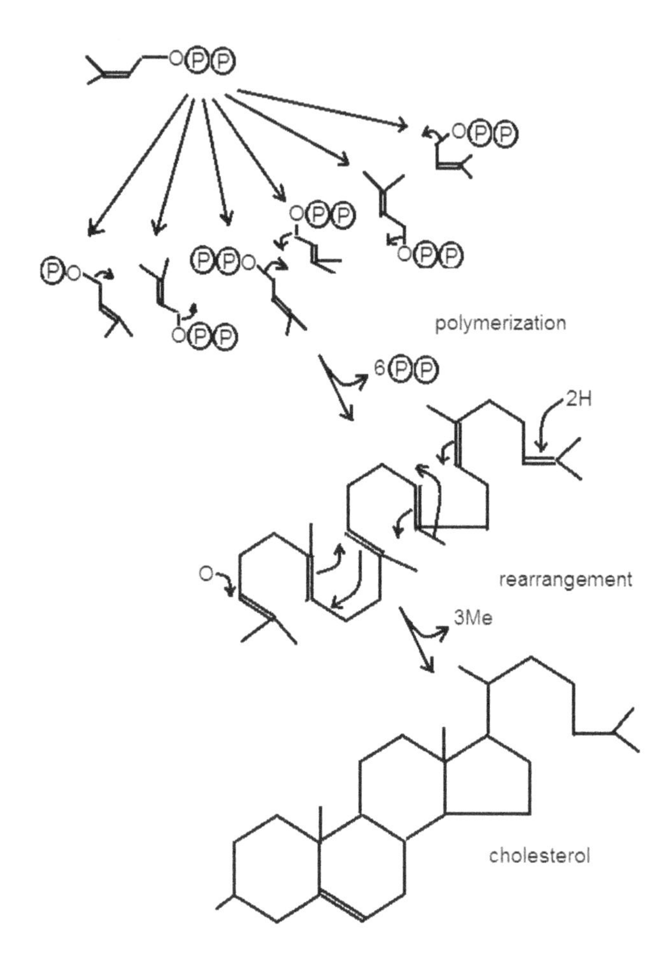

Fig 5.23 Cholesterol synthesis

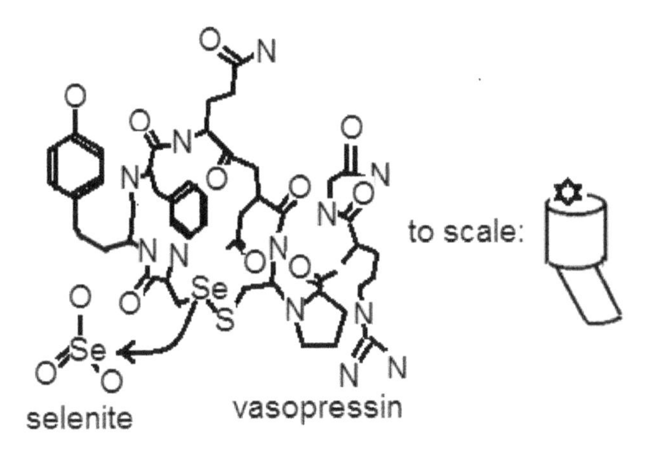

Fig 5.24 Oxytocin & vasopressin release selenite

Cholesterol is incorporated to cell membranes and used to build steroid hormones. Excess cholesterol blocks arteries, causing heart attacks and strokes. Prostaglandins produced by the prostate gland may provide osmotic support for sperm, egg and foetus during reproduction.

SeO_3= pumps Mn^{++}, *c.f.* sulphite, SO_3 = pumping Mg^{++}. The posterior pituitary packs selenium into hormones vasopressin and oxytocin for delivery. Vitamin E releases SeO_3=, *c.f.* glutathione releasing SO_3=, see section 5.1, Fig 5.24.

Changing consumption of cholesterol, saturated fats or Vitamin E has little effect. Mn nodules found on the sea floor suggest life has always had difficulties with this system. Fossilised animals deposit Se in cretaceous rocks; the hard water flowing from springs contains more Se than soft water from older rocks. Food processing reduces Se availability. Although water is odourless, it purveys odours; both hydrogen sulphide and hydrogen selenide, H_2S and H_2Se smell bad. Every SeO_3= changes Mn^{++} concentration significantly.

Pathology

Methyl mercury, $HgMe^+$ mimics trimethyl-selenium, $SeMe_3^+$, poisoning the system. The sulphur, S in super-phosphate fertilizer displaces Se, causing swayback in sheep. Se deficiency leaves us liable to heart attacks, strokes, hypertension during pregnancy and cancers of breast, colon and prostate, tissues handling water, Fig 5.25.

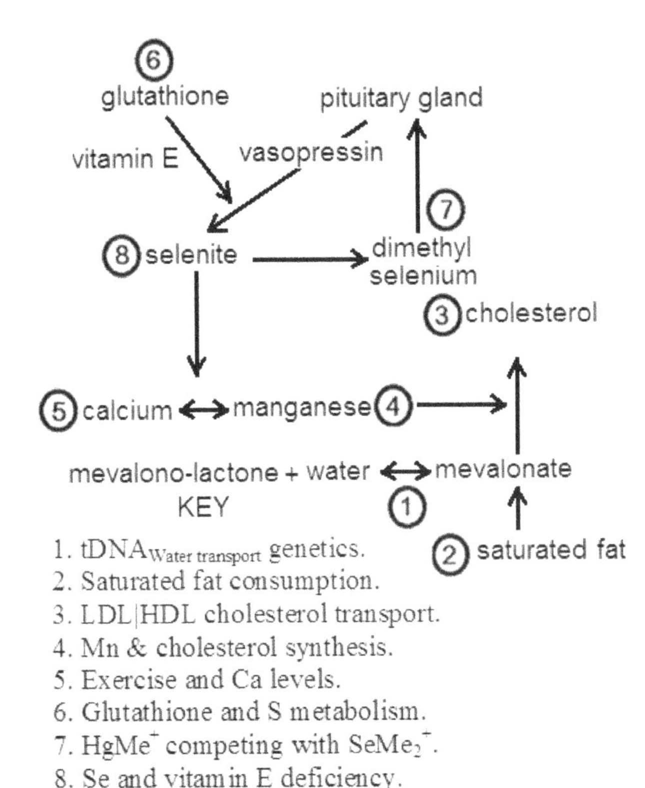

Fig 5.25 Controlling blood pressure

Emotions, drive and determination rule the heart, benefitting from practising hobbies, recreations and outside interests, not behaving like a cog in a machine. Over-excitement strains blood pressure, disrupted by Se deficiency. Complex and expensive pharmaceuticals, surgical operations and other treatments for high blood pressure have side effects. Promoting Se supplements could counter the pandemic of deaths from Se deficiency.

6 Evolution

Evolution is the achievement of progress, as incomprehensible as beauty but mental, not physical. Progress is change, it has no intrinsic value, whether changes are right or wrong calls for judgement. To be effective, progress must be planned, this book presents a plan for progress. Democracy depends on reviewing plans and discussing ideas; evolution consists in imagining options and applying accumulated knowledge to researching, developing and implementing them.

Imagination inspires hope for the future, not threatening social stability as a cultural or scientific revolution would but inviting realization of dreams consistent with prophecies and experience. I don't invoke astrology and numerology to predict the future but to prepare mentally as science prepares physically. Evolution can be disconcerting at first, but reasoned debate excludes weak proposals, allays fears and increases understanding.

As I wrote this book forty years ago, my ideas changed from vague proposals to detailed plans, methodical research hasn't revealed any reason to delay their implementation. Should flaws in my arguments emerge, I'll try to correct them, introducing changes can be difficult.

6.1 Change

Change is inevitable, either evolutionary, consequent on our actions, *acts of God* or resulting from human enterprise. To make changes, beliefs and understanding need be balanced. Biology describes evolutionary changes, *e.g.* how mushrooms and men replaced plankton and dinosaurs. Newspapers report accidental changes, the Bible, Koran and other holy books recount acts of God; some entrepreneurs create history.

Scientists seek to explain events and their consequences. Evolution and progress are unpredictable, but can be guided by reference to past experience. Apart from the fictitious names, the stories in sections 6.2, 6.5 and 6.8 are true.

After a carpenter fells a tree, he cuts it into planks and waits for them to 'season', then draws furniture designs to show to prospective customers. When the wood's ready, he cuts, smoothes and joins pieces then stains, paints or polishes the completed product. Satisfied customers reward him. Change has occurred, but no progress or evolution has taken place. I'll try using my experience to demonstrate how evolution works.

6.2 Time

New Year's Day was bright with a touch of frost. Most people stayed by their firesides recovering from festivities. Adel and I visited Windsor Great Park and walked to Windsor. The dew-spangled grass, miasma across the lake and gaunt forests of trees contrasted with cosy, noisy London. Time stood still, our minds untroubled by anything save the crack of ice on a puddle or the flight of a bird, free to wander.

Following a study of astrology and numerology, I'd concluded that ancient wisdom reflected common truths, the key lying in numerology. As we walked, Adel checked my logic. In *The Sign of Four*, Conan Doyle's Sherlock Holmes remarks *How often have I said to you that when you have eliminated the impossible, whatever remains, however improbable, must be the truth?*

Truth isn't an extract, but a combination of facts. Our conversation suggested that numerology doesn't overlay the mind, but underpins it. That idea condensed in our minds as the frost crystallized on the grass at our feet. Fig 6.1 shows the numerological birth charts I made for myself, Adel and the time we walked that afternoon:

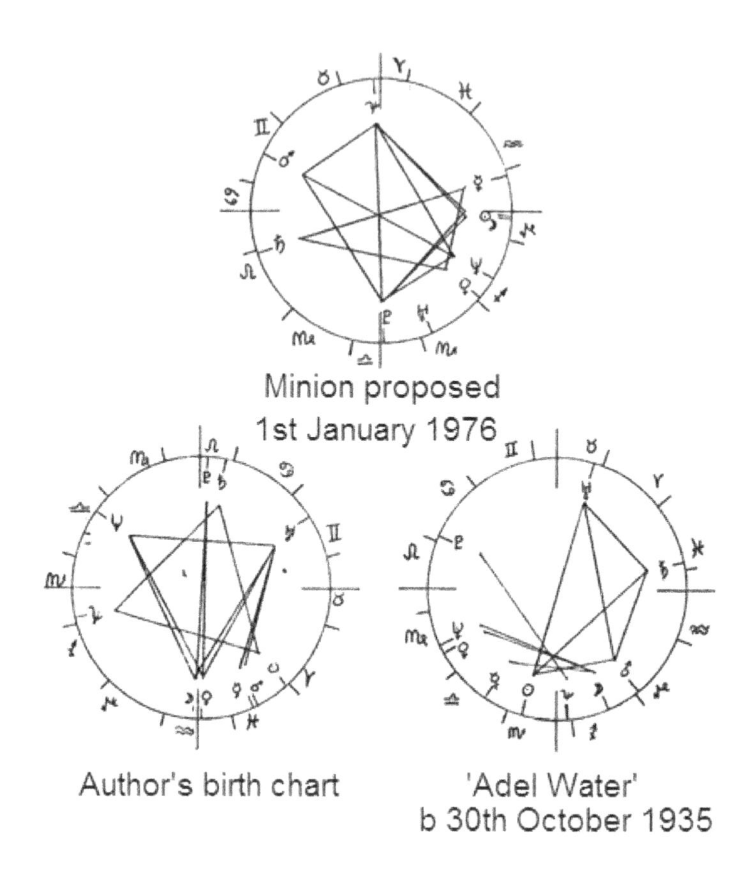

Fig 6.1 Astrological chart for minion proposal

Like thought itself, their symmetry is crystalline. By taking a rational view of numerology, we discovered a way of believing, determining that life has nine parts. As an illustration, this book has nine chapters, each with nine parts. If I believe in God, He has nine balanced parts and the New Age will establish a balance amongst them.

6.3 Survival of the fittest?

Our survival is inevitable according to science. A species avoids extinction by adapting its environment to enhance its survival.

Nothing is inherently unfit, Bengal tigers are natural predators without the endurance to find a meal on the savannah. Elephants have few predators, but can freeze to death in the Arctic. It's impossible to guess what the future environment will be like and equally difficult to judge species' potential to survive. One now at the top of the food chain may soon find itself at the bottom.

The phrase *survival of the fittest* is emotionally charged, fitness to survive isn't based on religious or political belief or health. Some believe that flouting these qualities renders a person unfit. Zealots in the past declared their version of the truth to be self-evident, deeming others unfit for their different beliefs. Believers in novel ideas need seek empirical tests to support them, nothing is self-evident.

Humans created histories, they're open to interpretation, not absolute truths. The original reporter's understanding was limited, there's always room for doubt. False evidence leads to false conclusions, scientific evidence can be misunderstood. History needs regular review, the version we teach should reflect the latest understanding. Factions of the population have different beliefs regarding their origins and fitness to survive.

By restructuring history, factions can be united without destroying their beliefs. Ancient textbooks and artworks needn't be destroyed, they just need reinterpretation. The best will survive, the rest be recycled. By recognizing information needing correction, our ideas change, allowing better resource utilisation. Outdated ideas need replacing by inventing new ones.

There's no limit to evolution and seeing the bright side of life, by thinking positively about the past we can envisage a better future. Remorse and recrimination over past mistakes leads to more of the same. Optimists lead happier lives; if our dreams crumble we can weave new hopes for the future, not to recreate former glory but to create new splendours. Believing in one's fitness and right to survive is important. We'll all be happy if righteousness survives and evil dies.

6.4 Choice

The nine ways of thinking introduced in Part 1, once understood, apply to every situation, Fig 6.2.

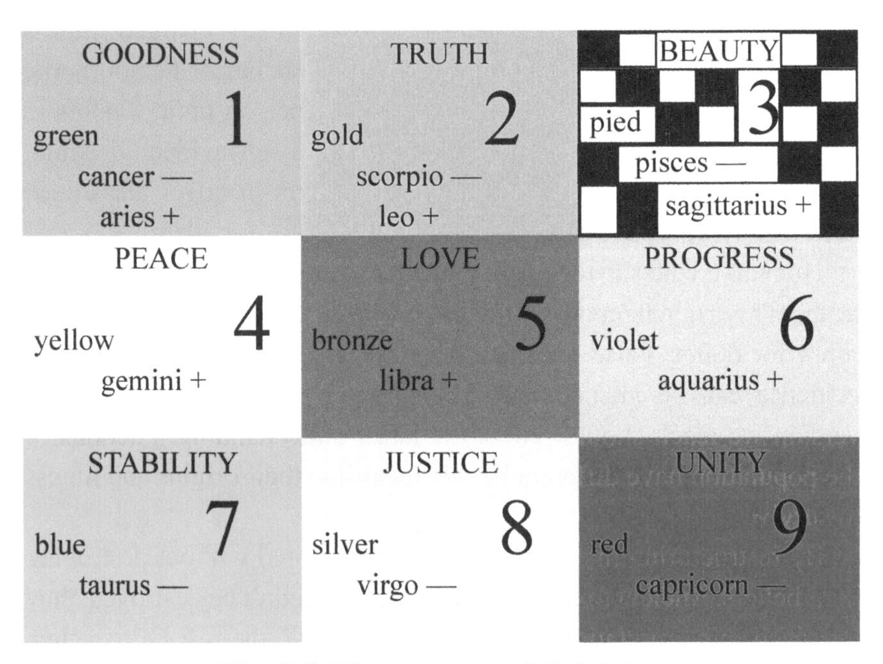

Fig 6.2 Nine ways of thinking

If any of the nine ways is difficult to understand, try meeting people born under the associated zodiac sign who think that way and look at the lists in tables 1.2–1.4 and 2.4. Choose twelve good examples to follow, each will offer a different angle on life and provide different advice. Consider their opinions and face reality, the nine ways used correctly can suggest solutions to every problem.

Changing your mind is easier than revoking a decision. Every idea has physical, mental and spiritual aspects. Choice determines the rate of evolution, by asking the right questions, we can keep things the way they are or improve them. Wheels weren't invented until someone asked *'What if I build a wheel?'* Choices are determined by laws, governments, habits and traditions.

Decisions depend on our knowledge and obeying society's laws, traditions and conventions. Choices couldn't be made without them. Planning and daydreaming enable you to make good decisions. Once you've made a solid plan, stick to it – be accountable for your actions.

6.5 Perspective

One summer, having leave from work in the hospital laboratory to study at university, I indulged my fancies, ran barefoot in the park, researched in the library and wrote this book. Postponing exam revision, I spent six weeks writing, whilst exchanging visits with Jean. That version didn't describe active transport; being invited to prepare a lecture on copper and zinc, sections 5.5 and 5.7 emerged, describing how those two elements function in metabolism. Nobody commented on my talk, my ideas were ignored; neither my friends at Cambridge nor an expert in Oxford responded to accounts of my discovery.

My life became a struggle, losing the will to eat or sleep I became delirious and unable to explain myself. The following ten days will always be remembered: visiting Blenheim Palace, a medieval play in Chichester, witnessing a police raid in Soho, being arrested at London Airport for choosing sweets from a closed shop and taken by Black Maria to a padded cell. I rejoiced in the view of freedom beyond barbed wire and exercising with other sick prisoners.

After an appearance in court, I ate any food gladly, sleeping on the prison floor until referred to a mental hospital. I feel sorrow, not bitterness recalling those events. They taught me lessons about human love and understanding others may not learn in a lifetime. I rejoice with the perspective of time, as Jean did when visiting me. Fig 6.3 shows the star patterns relating to that time:

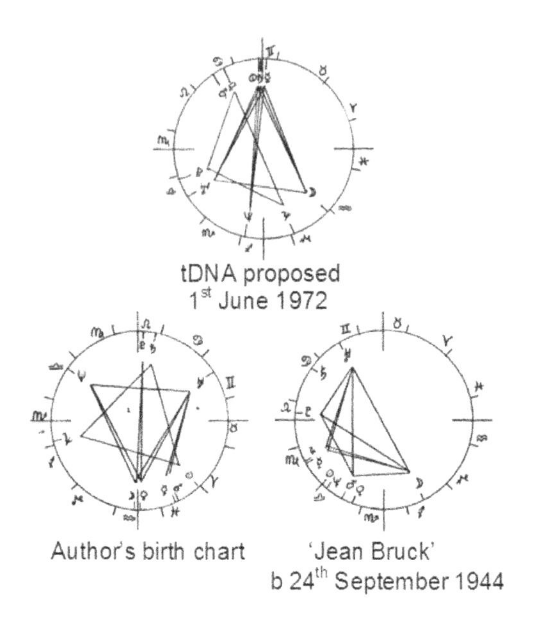

Fig 6.3 Astrological chart for tDNA proposal

That edition of this book was discarded, when I resumed five years later, my perspective had changed; this edition may need revision.

6.6 Evolution of thought

From Stanley D Beck's *The simplicity of science:*

Over a half century ago, Edwin A. Abbott, an English minister, published a fantasy entitled Flatland, based on an intriguing idea. Flatland was a two-dimensional world inhabited by two-dimensional people, Flatlanders. The world of our common experience is three-dimensional; things have length, breadth, and depth. In Flatland, however, there was no depth. People and objects had length and width only. Their world was simply a surface, like a sheet of paper, and the Flatlanders were figures, such as triangles and circles, drawn on the surface.

Flatlanders had no idea whatever of a third dimension, no inkling of up and down. Their movements, their experiences, and their thoughts were limited to the two dimensions for their peculiar existence. They encountered problems their two-dimensional scientists could not solve. The origin of light was such a problem. They found light in their world, but they could not tell where it originated, since they had no conception of space in more than two dimensions.

One of the Flatlanders, the hero of the fantasy, experienced a visitation or revelation, in which he was granted knowledge of three dimensions. In his vision, he visited a world of three dimensions. To him, this was a glorious and transcendental experience and he determined to tell his fellow Flatlanders of it. The Gospel of Three Dimensions was to be a gospel of a new world, a different world, more glorious and meaningful than Flatland.

This new world was their Flatland plus a dimension outside of their experience, knowable to them only by revelation. So inspired, the little Flatlander set out to evangelize all of Flatland. In order to explain the third dimension to others, he had to use their language and to express himself in their concepts and within their world of experience. This he found he could not do. He could not make himself understood; his inspiration was not intelligible to Flatland logic.

Eventually put in to an asylum for the incurably insane, he continued to write and dream of the world of the New Dimension. But as time went on, he had difficulty in separating his own mind from Flatland practicality, and he encountered more and more difficulty in keeping the memory of his revelation sharp and clear.

Commentators hailed the story as prophesying Einstein's relativity, giving Abbot more credit than was his due. He meant that some aspects of reality can only be known in part; the deepest insights and highest visions in the arts, religion and science challenge available means of communication. The existence of a third dimension couldn't be proved, so Flatlanders rejected it as nonsense and refused to believe it.

They couldn't imagine any way to test the idea. For the same reason, many ignore intuitive knowledge and religious experience. Spiritual revelations and gut instincts aren't necessarily false, messages are easier to accept if common symbols and language are used to describe them. Without any concept of a third dimension, Flatlanders had no way to communicate it, protagonist were regarded as insane. Mathematical symbolism may convey thoughts which can't be related in words. When opportunities arise, new ideas can be communicated, changing today's flight of fancy to future accepted truth. People declaring ideas wrong without affording them thought are *Flatlander*s.

6.7 Competition

When people ask me, *'If you're so clever, why aren't you rich?'*, I reply that I'm smart enough to know how to spend money whenever I have any. I rely on my wits to care for tomorrow and achieve my goals, wealth is secondary. Those blessed with high intelligence, IQ have an advantage – geniuses understand things more easily.

Ideally, gifted folk would share their bounty with everyone, but taxing intelligence encourages a brain drain. If everyone relied on an intelligentsia to solve every problem, they'd become subservient to them; people should be helped to help themselves. I propose three ways specialists' skills could serve others, each using a 'Big Brother' computer controlled by the user. I hope they're an appealing alternative to George Orwell's *1984*.

MYCALL would select which broadcasts you receive and promote them, using your preferences selecting entertainment, news and topics without divulging anything to third parties. At the user's whim, the menu changes, only programmes matching your preferences are shown, everything else is kept for future use. MYCALL connects like-minded people, reducing the cost of advertising by directing advertisements to likely customers. When you tell MYCALL your preferences, it optimises your use of communication time.

LIFELIGHT adjusts ambient lighting, improving the quality of life by using CIS, Complementary, Independent and Supplementary modes. Sensing light sources, sounds, odours, the time, weather conditions and whatever else electronics can detect, it sets the lights to enhance your surroundings. Its continuous light therapy adds colour to daily activities, detects your feelings and adjusts the lights to balance your mood.

You glow with joy when you're happy, bringing others closer by expressing your feelings. LIFELIGHT can be made sensitive to music, bird song and conversation, compounding the effect of a storm, brightening a dull day or enhancing morale in a factory. It ensures you see things your way, expressing itself using a collection of tones significant to the listener.

LIPSOUND composes music from words, creates an accompaniment to a poem for a loved one or a lyrical advertisement. It could translate your essay into music which you can change by editing the text. You compose music using it, creating a symphony without knowing musical notation. Playing the product electronically won't surpass a live performance.

MYCALL, LIFELIGHT and LIPSOUND exploit minion logic, savvy entrepreneurs could improve them, hopefully not for military purposes. Experts will find them easy, everyone can gain some benefit and improve their quality of life.

My heroes, Isaac Newton, Charles Darwin, Michael Faraday and Linus Pauling all hoped life would prosper. A scientific education could enable children to advance that cause. Everyone has a part to play in the New Age, allowing no dictator or monopolist to be Big Brother, don't follow any leader blindly. New Age citizens aren't sheep, they're more like goats.

6.8 Dreams and reality

The summer before my second year at Cambridge I travelled ~1800 miles round Ireland by bicycle, earned a bit at a cosmetics firm, had an aerosol valve cap I invented considered for development and found friends for company. Following a good examination record it sent me high in the winter term. A bitter winter, loneliness and isolation and an unfavourable gender ratio left me depressed, lacking a goal in life. Having abandoned conventional religion, I sought solace with the humanists and physical science.

I spent long nights exploring classical physics, seeking something more concrete than the paradigms Einstein disliked. I hoped to emulate Michael Faraday, the lovable scientist and surpass his contributions in depth and scope, feeling my fortunate background would help. While the Sun was in Aquarius, from late January to early February, I attended a poetry reading, *Thoughts in Solitude* whilst conducting a physics practical on the Clausius-Clapeyron relation for the boiling point of nitrogen.

William Butler Yeats wrote *What Then*:

His chosen comrades thought at school
He must grow a famous man;
He thought the same and lived by rule,
All his twenties crammed with toil;
'What then?' sang Plato's ghost. 'What then?'

Everything he wrote was read,
After certain years he won
Sufficient money for his need,
Friends that have been friends indeed;
'What then?' sang Plato's ghost. 'What then?'

All his happier dreams came true --
A small old house, wife, daughter, son,
Grounds where plum and cabbage grew,
poets and Wits about him drew;
'What then.?' sang Plato's ghost. 'What then?'

The work is done,' grown old he thought,
'According to my boyish plan;
Let the fools rage, I swerved in naught,
Something to perfection brought';
But louder sang that ghost, 'What then?'

I also read his *The Shadowy Waters* and *The Choice*:

It is a love I am seeking for,
But of a beautiful unheard-of kind that is not in the world.

During two weeks of mounting excitement I tried every possible explanation for my surprising results, finally believing I'd made a discovery. Fig 6.4 shows the associated time pattern, it bears similarities with Figs 6.1 and 6.3; all chime with my nativity.

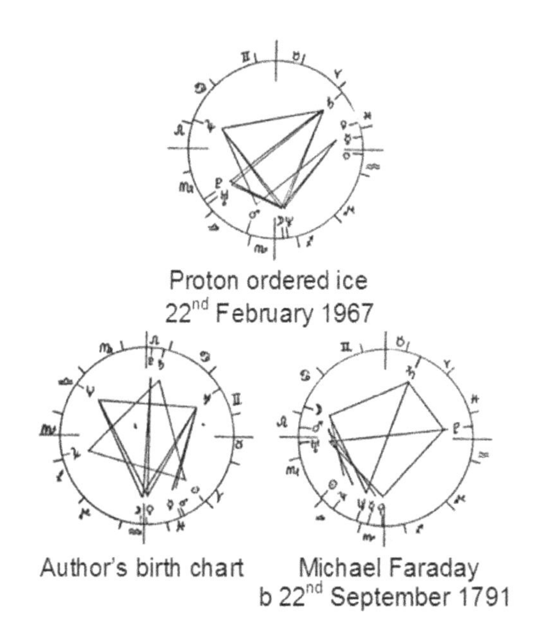

Fig 6.4 Astrological chart for ice It discovery

My over-excitement had me committed to mental hospital, starting the systematic destruction of my career aspirations and years of ridicule. My discovery has hurt my family and friends and upsets the scientific status quo. These are part of the cost of realizing a dream, it's difficult to reconcile them with history, to see them as evolutionary – every dream has its cost. Don't let the price deter you, no one will encourage you – realizing your dream is up to you.

6.9 Social evolution

The beast

The history of mankind records tales of people fighting to realize their wide-ranging dreams, with many goals and methods. Aleister Crowley, the English poet and occultist had dreams which he took very seriously and recorded in *The Book of the Law*, hoping to witness the dawn of the New Age. Paragraph 3:47 says:

This book shall be translated into all tongues: but always with the original in the writing of the Beast; for in the chance shape of the letters and their position to one another: in these are mysteries that no Beast shall divine. Let him not seek to try: but one cometh after him, whence I say not, who shall discover the Key of it all. Then this line drawn is a key: then this circle squared in its failure is a key also. And, Abrahadabra, it shall be his child & that strangely. Let him not seek after this; for thereby alone can he fall from it.

Crowley reckoned himself to be the Beast prophesised by John the Baptist in the biblical Revelations 13:18:

Here is wisdom. Let him that hath understanding count the number of the beast: for it is the number of a man; and his number is Six hundred threescore and six.

Crowley's numerological number is 69669, see section 6.5. His book *Liber Aleph 111* (based on *The Book of the Law*), subtitled *'The Book of Wisdom or Folly, in the form of an Epistle of 666 the Great Wild Beast to his son 777'*, was published in 1960 in *The Equinox* volume 3.6. Crowley died 1st December 1947, disappointed not to have identified 'his child' *Liber Aleph 111*.

Like Revelations and his Book of the Law, it's difficult to read. It reflected his thoughts on people who would keep magical New Age secrets. He wasn't alone in trying to establish an elite ruling class, with terrible consequences. A secret society, however benevolent its intentions, can't be trusted. The star pattern relating to the time Crowley wrote his book may suggest why he attached importance to it, Fig 6.5.

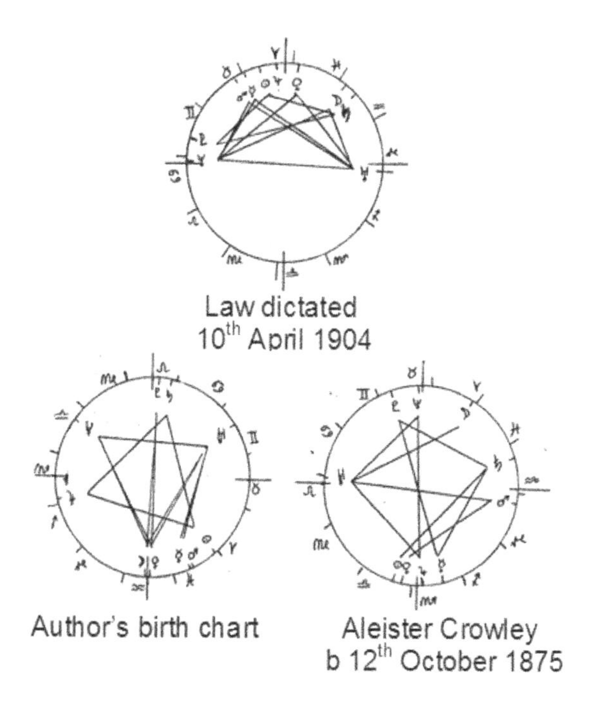

Fig 6.5 Astrological chart for Crowley's writing

Historians now treat social evolution as more significant than the rise and fall of kings. The need for philosopher kings has diminished as Plato's various forms of society have developed. The New Age doesn't need a leader with secrets or an appointed starting time. Everyone is free to declare that it's begun and include everyone as members. It begins when we're ready and will end when the Age of Capricorn's time comes around.

I don't dismiss the insights afforded by prophets, magicians or occult tradition. I refuse to accept a particular interpretation of or deny any individual's importance in social evolution. Try reading Crowley's and John's writings, they reflect crystallizations of thought occurring at special times. The people in whose minds those crystallizations occurred are unimportant. Distortions in their writings can be corrected by knowing about them. Societies evolve by studying inherited information. Social evolution is delayed if it's suppressed, censored or kept secret. All worthy dreams will come to fruition in good time, dreamers should report their dreams openly and let others judge them.

Conclusion: Life

I proposed in the introduction to study life, determine what's natural and how we could lead a balanced existence. Life is a dynamic equilibrium, a three-way balance in motion, nature doesn't redress imbalance but adjusts to change. Nature makes no distinction between changes fostering life and those restricting it, every person need exploit changes to live a natural, balanced life.

We must use resources and dispose of waste carefully, using all the nine ways and keeping our bodies in good order. When changes are needed, we must organize them well. We best try to live truly natural lives free of restrictions, enjoying New Age freedoms and hopes; a final quote from Aleister Crowley:

Do what thou wilt shall be the whole of the law. Love is the law, love under will. There is no law beyond do what thou wilt.

Part 3: Origin of Life

Preface

This Part indulges the freedom from stability, justice and unity afforded by the New Age, my proposals, based on a chance observation at Cambridge in 1967, need empirical testing. In 1974, a journal editor wrote: "These are purely speculative suggestions which are not based on established data, or the existing body of biological theory". It completes the picture of the beginning of life on Earth by adding to the ideas presented in Parts 1 and 2. By merging research from many disciplines, it counters a tendency to compartmentalise scientific knowledge.

I've researched these proposals for over forty years, embracing as many fields of learning as possible. They should enable students to regain Aristotle's status of universal understanding and pave the way for a model with yet greater scope. Arising at the dawn of the New Age, everyone is responsible for nurturing them. My amendments to scientific laws evolved from their predecessors. I'd prefer them to be termed New Relativity, New cosmology, etc, not using my name. I delight in them as revelations, like an explorer describing a new land. I propose a moratorium on expensive research on obsolete theories of particle physics and astronomy until the world's real problems are resolved. This is my only plea, I seek to unify science, not change the course of history.

7 Molecular life

The fossil record supports Charles Darwin's contention in *Origin of species* that life on Earth originated in a sea of chemicals. This chapter uses the same logic to explain life's origins, describing how life would recur if wars or other disasters extinguished it. I use some scientific terminology as in parts 1 and 2. School curricula should give science the same importance as reading, writing, arithmetic and technical skills and science teachers understand my book. We teach nothing if believe nothing, passing unresolved mysteries on for them to resolve. I hope society will use my ideas for its benefit.

7.1 Primordial soup

Life's simplest imaginable starting point is a desert with oases. ATP and the elements and compounds introduced in Part 2 are available, chapter 8 discusses their origins. This chapter concerns selection and multiplication of particular molecules.

The primordial soup is in dynamic equilibrium with potential for life. Atoms and molecules move about rapidly at random, they're no more confused than a termites' nest. We can't say in the absence of clocks when order arose out of chaos. My thought experiment addresses a timeless problem, time stands still until something happens.

7.2 Ice, the ordering force

Ice-light

Initially, nuclear reactions contributed the only order, leaving parts of the plane hot and others cold. They caused water to evaporate in warm areas and freeze in cold ones. Nitrogen liquefied in the extreme cold. Rain, hail or snow falling into puddles of liquid nitrogen made it boil, carrying water vapour in the bubbles. It crystallized, producing crystals of ice It, the proton ordered ice I discovered at university. These crystals share diamond's orderly structure and the beauty of snowflakes. Being ferroelectric, polarised like magnets, they glow in the dark, Fig 7.1.

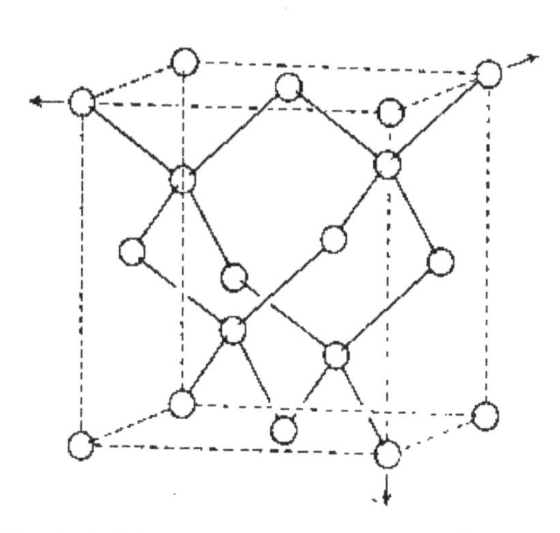

Fig 7.1 Diamond structure of ice It

Ice It crystals undergo a phase transition on cooling which accommodates the water molecules' irregular shape, releasing flashes of infrared laser ice-light like an army marching in step destroying a bridge, Figs 7.2 and 7.3.

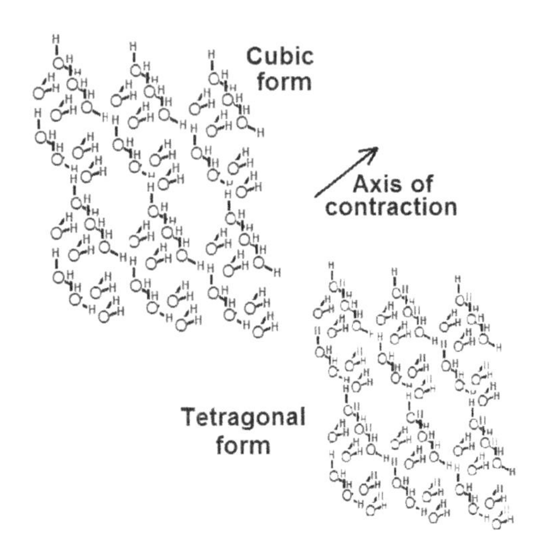

Fig 7.2 Ice It ferroelectric transition in perspective

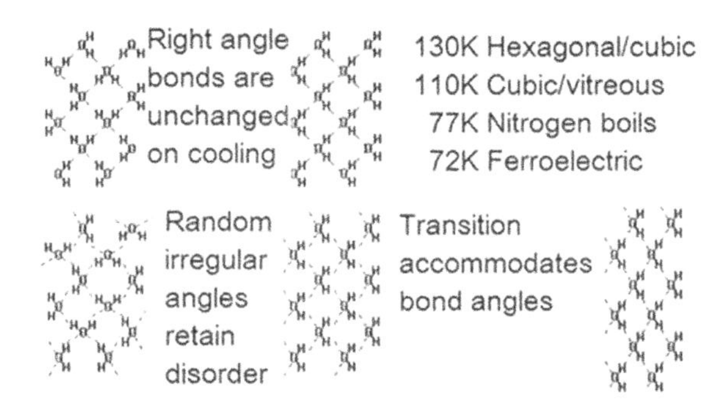

Fig 7.3 Ice transition in 2D and temperatures

Concentrating

Ice in clouds and on the plane surface reflected ice-light like the radio signals Guglielmo Marconi sent across the Atlantic and polarising it. Fig 7.4 shows ice-light reaching the primordial soup.

Molecules absorb light of different colours, short wave ultraviolet, UV and long wave infrared, ir are invisible. Ice-light excites and phosphorylates such deoxyribodinucleotide molecules as dADP like an opera singer setting a wine glass vibrating. They form such deoxyribotrinucleotides as dATP, dynamic equilibrium replenishes the deoxyribodinucleotides. Since the light is polarised, only left- or right-handed versions of DNA result; life is 'chiral'.

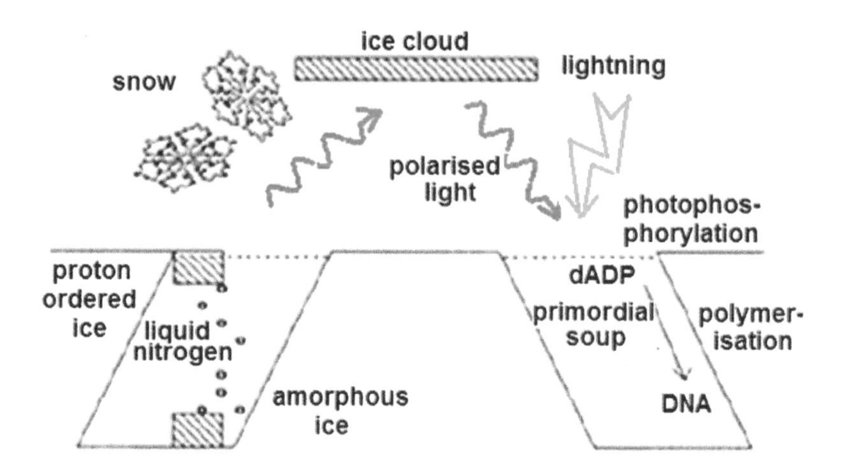

Fig 7.4 Origin of life: Ice It and 4μ infrared laser light synthesize DNA

7.3 Pumps and cells

Cells, noodle soup and pumps

Fat molecules form films on the water surface with their water-loving, hydrophilic ends down and water-hating, hydrophobic ends up like detergent molecules in a washing-up bowl. Gusts of wind form bubbles, the primitive cells, *coacervates* Alexander Oparin proposed, folding them double. When volcanoes or nuclear explosions evaporate the water, the deoxyribonucleotides are concentrated, forming a noodle soup of DNA, DNA is more stable than RNA. Some DNA forms transport DNAs, tDNAs which embed in coacervates – a chicken has laid an egg. Life has started.

First life, alien life and genetic engineering

A flash of lightning charges the cell and tDNA pumps feed it until it bursts, forming two bubbles – it's reproduced. Cells provided with a varied diet by several tDNAs compete for substrates more successfully. Those accumulating DNA's components make tDNA.

An extreme ice age, a primordial soup surface, heat concentrating dATP and a variety of tDNAs are prerequisite to life forming, it could occur on any Earth-like planet. If another natural laser synthesised another self-replicating polymer, life with different chemistry could arise. No such regime is known, ice-light and DNA may be a unique combination. Genetic engineers may someday create life artificially.

7.4 Reproduction

Diversity

Base-pairing enables DNA to replicate, cells with useful tDNA combinations have an evolutionary advantage. The most efficient species consumes everything until some change occurs. Thomas Malthus argued that population increases are limited by the means of subsistence. Proliferation locks all resources in the most efficient life form. The world's expanding population threatens to exceed natural resources.

Proteins and efficiency

Chemical reactions run slowly without catalysts, trace metal ion complexes speeded the prerequisite concentration of life's ingredients. Ribosomal RNA, facilitating protein synthesis and the enzymes generated came later. Polypeptides countered DNA's acidity and prevented it from forming a double helix, as in the minion chip in the brain introduced in Part 1, later serving as hooks binding cells together.

Cells producing energy without ice-light have an advantage. The porphyrin molecules found in red haemoglobin, green chlorophyll and cytochromes convert sunlight and energy released by digesting sugars to 4μ infrared ice-light, see Fig 5.1. Life generates energy in many ways: it survives in deep rock strata, hot springs, arctic ice and the ocean depths.

7.5 Diversification

Mutation

Life relies on random mutations to advance. Mutant offspring compete with their parents, forming new species. Most mutations are changes in DNA encoding proteins; rare familial diseases evidence mutant tDNAs. Evolution involves either improving an existing function or inventing a new one. *Sitting thinking* or *green-field* research may yield useful ideas, *e.g.* for disposing of nuclear waste. Mutations create biochemical pathways possibly proving useful to their offspring.

Survival

The simplest version of survival of the fittest is outstripping opponents in exploiting scarce resources. Mental evolution is exemplified by technological advances and spiritual evolution by inter-religious strife. Recessive traits are transmitted to offspring but not translated, manifest only if both parents carry the same recessive gene, affecting their progeny in some way.

There are thousands of cripples for every champion – uniform populations are poor substitutes for excellence. 20th century totalitarian governments attempted eugenics, purging populations of undesirable genes by selecting desirable traits. If genetic engineers select mutations, they can bypass nature, many consider it unethical. To balance life, changes imagined at leisure are needed. I next consider how molecular evolution led to modern life.

7.6 Molecular ecology

The language of life

Life becomes complicated when survival of the physically fit distorts its balance – brute strength yields to subtlety. No individual in the primordial molecular ecosystem thrives in isolation, species support one another. Scarcity of resources imposes constraints. Every element has a role, either in active transport or growth. Elements are the alphabet of the language of life, spelling every word.

Life still uses its original atomic and molecular alphabets, introduced in part 2. Global solutions for ecological problems are confined by its rules – heroic measures to rectify climate change disregarding them could prove disastrous. The disadvantage of assigning functions to elements rests on scarcity – too few Zs would render a typesetter's task when printing a geometry book difficult.

Progress

Selenium deficiency, causing common cancers and blood pressure disorders is an important case. Substitutes can help, *e.g.* managing bipolar disorder by replacing iodine with lithium or treating Parkinson's disease with synthetic drugs like $_L$Dopa. We can distribute chemicals and fertilizers, add supplements to our food, introduce species to new habitats or genetically engineer others. Eventually, a new atomic alphabet may be invented.

Dangers of modifying tDNA

Life's survival could be assured if decision makers in pharmacy, agriculture, veterinary chemistry, cosmetics and food processing had a broader education. Pure chemicals are invariably contaminated. Manufacturers handling toxic chemicals must dispose of waste carefully.

Fertilizers must replace elements crops remove from the soil. Cigarette smokers and alcohol drinkers have endangered their health for pleasure. Objecting to using regulated toxins in food production or disease prevention is pointless. Molecular ecosystems should be adjusted with caution, novelty can be dangerous.

Rewriting DNA sequences encoding proteins has predictable consequences. Designing novel tDNAs could have unforeseen results – proteins containing fancy $\alpha\alpha$s might prove indestructible, viruses invading cells through them impossible to control.

7.7 Biochemistry

ATP, the middleman

ATP delivers the energy which ice-light originally provided, the cytochrome chain recharges it via 4μ-size parcels created by porphyrins, see Fig 5.10. Sources include mitochondria digesting sugar and chloroplasts absorbing sunlight, both trap 4μ energy in membrane envelopes, replacing primordial ice-light.

All plants and animals have similar mitochondria, cellular power houses. Biology uses a limited range of ideas, applied in ingenious ways. Research on experimental animals is useful. Experiments with mice or monkeys can test cures without distressing them if thoughtfully designed. Ethical standards aren't absolute, objection by tree-hugging animal lovers aren't helpful.

Advances

Recent advances merely scratch nature's surface, accommodating them will take centuries, few changes are completed in a lifetime. Dreamers imagine overnight transformations, discoverers and inventors are recognized posthumously, their contributions take time to disseminate. Prophets and visionaries prepare paths for smooth transitions. Observing running deer, flying albatross and swimming fish prepared the way for cars, planes and ships, enabling us to share their abilities.

7.8 Mind and muscle

Energy conversion tricks have evolved for various purposes using resonant cavities resembling mitochondria and chloroplasts. Entropy relates energy and order, *c.f.* light energy arising from ordered ice. Entropy, disorder, is measured in degrees like temperature, there's no absolute criterion, see Figs 7.5 and 7.6.

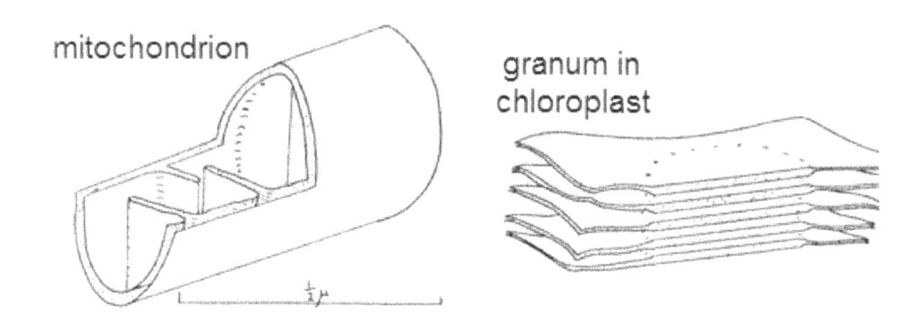

Fig 7.5 Mitochondria and grana are commensurate with 4μ in size, resonating with 'ice light'

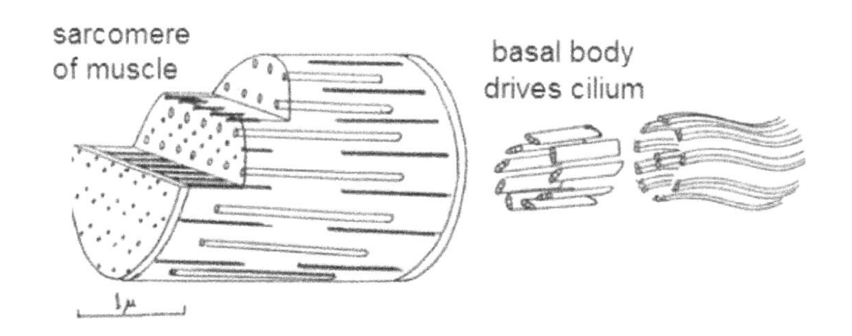

Fig 7.6 Sarcomeres and cilia are further examples

Muscle

Sarcomeres, units of muscle, are ½ an ice-light wavelength long. Energy released by ATP breakdown resonates in mitochondria, chloroplasts and sarcomeres. Activating a muscle causes its component fibres to contract to accommodate ice-light, no making and breaking connections between actin and myosin proteins is involved.

The whiplash action of cilia, hairs on the lung surface and other biological systems uses a base-nine counter, *c.f.* minion counting in 4.5. Like sarcomeres, basal bodies hold standing waves of ice-light. Fig 7.7 shows how they power chromosome separation. These energy couplings are more efficient than heat engines.

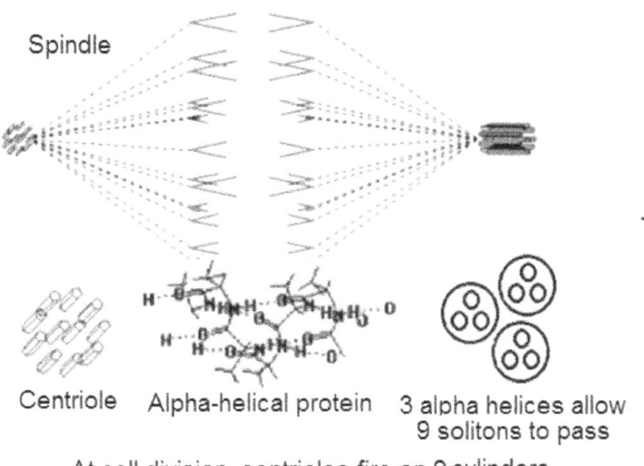

At cell division, centrioles fire on 9 cylinders, sending energy 'solitons' along 9 track spindle proteins. The chromosomes become magnets, pushing one another apart.

Fig 7.8 Chromosomes at cell division

Cold fusion

Biology extracts useful energy from every quantum. Minions, the mind components introduced in 1.2, have tunnels flanked by oscillating H-bonds accelerating protons, H^+ along them, Fig 7.8. Some fuse with carbon or nitrogen nuclei, driving the carbon-nitrogen cycle, *c.f.* Fleischmann and Pons's report of cold fusion in similar tunnels on the surface of palladium, Pd. Hydrogen is converted to helium, releasing energy as γ-rays at ambient temperatures, without the hazards of plasma fusion. If the energy could be trapped, they'd provide clean power, bypassing the Sun. Chapter 8 presents nuclear forms and their physics.

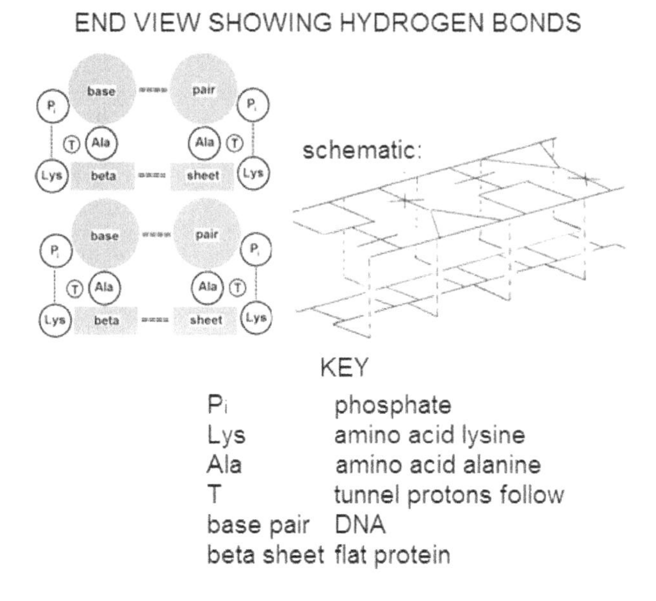

Fig 7.7 Minion tunnels, T accelerate protons fast enough for cold fusion

7.9 Photosynthesis

Section 5.9 described water pumping in animals. Plants' cellulose cell walls retain water, freeing water pumping slots in their tDNA dictionaries for the photosynthesis and nitrogen fixation described in chapter 4. Plants and animals are symbiotic, depending on each other for survival. Photosynthesis replaces ice-light, converting sunlight's nuclear energy into chemical energy. Sheets of chlorophyll collect light and tDNA light pumps using molybdenum, Mo transfer it across the cell membrane, Fig 7.9. Mo's nucleus is large, of all bioactive elements only silver and iodine are larger.

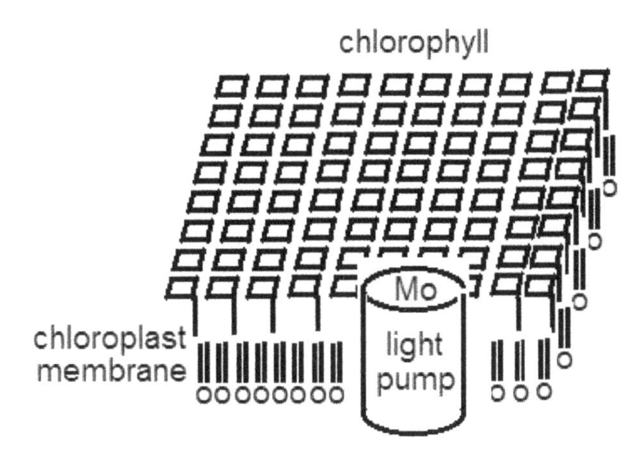

Fig 7.9 Molybdenum and chlorophyll collect sunlight for photosynthesis

Most photosynthetic energy is used making cell walls to resist osmotic pressure. Constructing nuclear fusion reactors is also energy intensive. Life is restricted to nine parallel activities, plants pump light and animals pump water, little green men doing both aren't possible. This system is self-sufficient apart from the Sun converting nucleohistone cosmic rays into light. Creating life in a chemical soup is only the beginning of the story, if atoms were made in minion tunnels, what was there originally?

8 In the beginning

The Bible recounts the beginning in Genesis chapter 1:

1 *In the beginning God created the heaven and the Earth.*
2 *And the Earth was without form, and void; and darkness was upon the face of the deep. And the Spirit of God moved upon the face of the waters.*
3 *And God said, 'Let there be light'; and there was light.*
4 *And God saw the light, that is was good: and God divided the light from the darkness.*
5 *And God called the light Day, and the darkness he called Night. And the evening and the morning were the first day.*

St John chapter 1:

1 *In the beginning was the Word, and the Word was with God, and the Word was God.*
2 *The same was in the beginning with God.*
3 *All things were made by him; and without him was not anything made that was made.*
4 *In him was life; and the life was the light of men.*
5 *And the light shineth in darkness; and the darkness comprehended it not.*

St John's 1st epistle chapter 4:

7 *Beloved, let us love one another: for love is of God; and every one that loveth is born of God, and knoweth God.*
8 *He that loveth not knoweth not God; for God is love.*

Sir Thomas Browne *The garden of Cyprus* chapter 5:

All things began in order, so shall they end, and so shall they begin again; according to the ordainer of order and mystical mathematics of the city of heaven.

This chapter recounts my mathematical axioms – scales, symmetries and infinities. It tries to declare, not explain, the meaning of love. My mathematical theorems are unproven. Axioms are judged by their efficiency in modelling reality, they aren't inherently right. We see the real world directly, its model 'through a glass, darkly'. Its mysteries may be revealed if we respect God's justice and love. The Lord's Prayer, omitting the words Love and God, becomes the *Loved's Prayer*:

Our teacher, which art in heathen,
Respected be thy message;
Thy kingdom come,
Thy will be done,
Universally, as it is in our hearts.
Give us this day our basic needs,
And forgive us our trespasses,
As we forgive them that trespass against us
And let us not think all well
'Til thine is the power and glory for ever.
As it was in the beginning,
Is now and ever shall be,
World without end.
For all men.

8.1 Relativity

Chapter 1 described minion counting, making wrap-around errors of 1 in 63^9 and 1 in 63^{18}. Newton's and Einstein's relativities assume light travels in straight lines, the shortest distance between two points. The minion sees straight lines as the curve obtained by integrating the equation, Fig 8.1:

$$\left.\begin{aligned}\frac{d^2\Theta}{dt^2} &= \frac{\pi\beta^2}{2\sqrt{e([1+\beta]^{t/\tau}+[1+\beta]^{-t/\tau})}} \\ \frac{d^2\Phi}{dt^2} &= \frac{\pi\beta^2}{2\sqrt{e([1+\beta]^{t/\tau}-[1+\beta]^{-t/\tau})}}\end{aligned}\right\}$$

Fig 8.1 *Tyger* time-gravity or relativity equation

It's named *Tyger* after William Blake's poem *The Tyger* from *Songs of Experience*:

Tyger Tyger, burning bright
In the forests of the night;
What immortal hand or eye,
Could frame thy fearful symmetry?

In what distant deeps or skies,
Burnt the fire of thine eyes?
On what wings dare he aspire?
What the hand, dare seize the fire?

And what shoulder, & what art,
Could twist the sinews of thy heart?
And when thy heart began to beat,
What dread hand? & what dread feet?

What the hammer? what the chain
In what furnace was thy brain?

What the anvil? what dread grasp,
Dare its deadly terrors clasp?

When the stars threw down their spears,
And water'd heaven with their tears:
Did he smile his work to see?
Did he who made the Lamb make thee?

Tyger Tyger burning bright,
In the forests of the night:
What immortal hand or eye,
Dare frame thy fearful symmetry?

A plane seen this way appears to have the same curvature as the Earth. Light leaving the plane approaches the vertical, appearing to return from distant surfaces the size of Sun and Moon. Like quantum mechanics, it's hard to accept, challenging the intuitive simplicity of Newtonian mechanics. The Tyger equation arises from minions miscounting by 1 in 63^9, the β-effect; errors of 1 in 63^{18}, the α-effect, also occur. The age of the universe always equals $\tau/\alpha \approx 10{,}200$ M years, suggesting we take it with us and we're effectively at the beginning.

8.2 Sun and stars

The cosmic rays released from minion tunnels appear to emanate from the Sun and an associated diffraction pattern, the stars; their arrangement reflects the minion's symmetry. Atomic nuclei travel like trucks in a railway tunnel, heavier trucks run slower, producing pulsars which decay as nuclei decay. Minions provide a useful basis for studying cosmology, implying there's nothing special 'out there', neither neutron stars nor intelligent life.

Sunspots waxing and waning reflect events on Earth. Mankind's survival is inseparable from the Sun's, it's at our mercy, world government might control it. The Sun obeys the 'cosmological principle', part of time-symmetry, representing the will of others. It's reasonable to ignore the possibility of a meteor colliding with the Earth or a black hole swallowing it. Precise astronomical measurements may teach us more about biology and how to construct a nuclear fusion reactor. The heavens reflect life, revere life, not its reflection!

8.3 Newton's apple

Cavendish measured the force causing things to fall to the ground, Newton's gravitational constant. The Tyger equation predicts that objects approach the velocity of light after falling for as many days as the number of years for the surface to resume its position relative to Earth and Sun. The surfaces implied are those of moons and planets; minion logic predicting their relationship supports astrology. Planetary positions as viewed from Earth are symmetrical, details of my cosmological model need refinement. Christopher Smart's *Song to David* v40:

> *'Tell them I am', Jehova said to Moses;*
> *while Earth heard in dread,*
> *And smitten to the heart,*
> *At once above, beneath and around,*
> *All nature, without voice or sound,*
> *Replied 'O Lord, Thou art'.*

Every large thing has a small counterpart, Tyger symmetry is the same above as below. I proposed in chapter 1 that atomic nuclei were composed of combinations of planes, so the largest stable nucleus would have 250 planes. Nuclei with the same number of planes using different shapes exhibit allotropy, Figs 8.2 and 8.3.

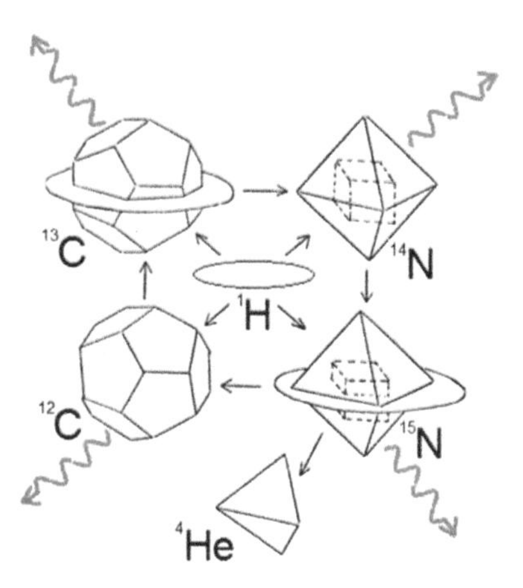

Fig 8.2 Carbon-nitrogen cold fusion cycle

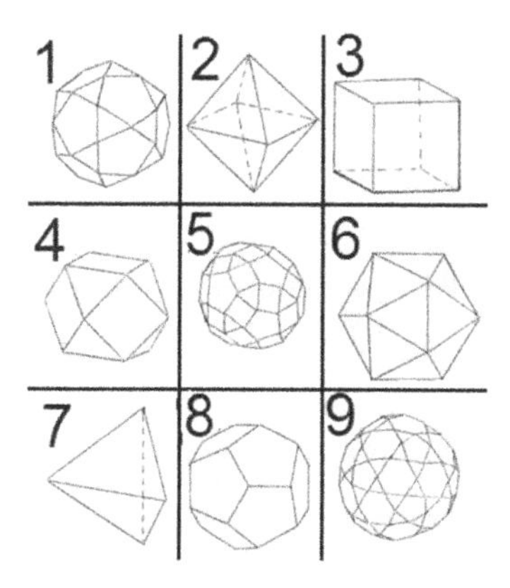

Fig 8.3 Nine 'perfect' solids explain periodic table

Einstein's $E = Mc^2$ suggests everything can be reduced to planes and energy; presumably one or other came first – a residual mystery of *the beginning*. Without energy or planes the world is inconceivable. I prefer to focus on the relationships between parts, not their origin. In *Memoirs of Newton II*, chapter 27, Brewster wrote:

I do not know what I may appear to the world, but to myself I seem to have been only a boy playing on the seashore, and diverting myself in now and then finding a smoother pebble or a prettier shell than ordinary, whilst the great ocean of truth lay all undiscovered before me.

8.4 Adam's apple

I've so far been concerned with good and of evil. Like Adam and Eve, you've consumed the fruit of the *Tree of Knowledge*, and Genesis declares you must die. The biblical serpent says in the Bible, Genesis 3:5:

5 For God doth know that in the day ye eat thereof, then your eyes shall be opened, and ye shall be as gods, knowing good and evil.

Like the serpent, I tempted you and you've eaten.

8.5 Maths and models

Although I've analyzed my ideas for many years and compared them with others', they stand alone. Mathematical logic enables deductions and distinctions using 'therefore', not inductions and combinations using 'because', it doesn't suggest reasons. This book is structured to illustrate base nine thinking. If my logic's wrong, the theory is incomplete; it may provide a stepping-stone to a better one in the future.

Debates can be lengthy, but there's virtue in economy of argument. If an equation according with nature can be substituted by a simpler one, it's probably wrong. To assess simplicity or efficiency, many factors come into play. Replacing Newtonian mechanics with the Tyger equation needs review. No formal religious belief satisfies everyone, readers need to accommodate base-nine logic within their existing philosophy.

8.6 Theoretical evolution

In May 1978, I resolved to write a sequel, *Peace Building* after *Science Uncoiled* was published, to guide politicians and diplomats in reconciling disparate nations, races and creeds. A third volume might identify ideas for scientists to research and artists to explore. To research *Peace Building*, I organized midsummer hilltop gatherings, took companions for a long-distance walk and sought entrepreneurs to develop my ideas on the minion. I proposed *Helicore,* a model of the mind and sought moral support, collaboration and publicity for my endeavours.

Both the global community and individuals could benefit by participating in midsummer peace rallies. Publication requires research, writing, editing to address a wide audience, distribution and delegating responsibility. Publications may be misinterpreted, misquoted and precipitate protest if ulterior motives are imputed. Mass-production devalues innovations. Inventors get blamed for the consequences of military applications; competition and plagiarism can arise.

Judgements often invoke laws introduced for other purposes. Folk with grievances may threaten innovators lives. Should they resign or take responsibility when damned for publishing? Faulty goods need repair, misunderstandings correcting, rules reinterpreting and objections debating. These are matters of conscience. Law courts defend rights and prosecute libel actions. Nobody happily resigns, goes bankrupt or accepts corruption.

8.7 Heaven and Earth

The powers and functions of leaders are governed by population size. The formula $63^N/2$, where N=1…9 suggests population groupings, Table 8.1.

Population	Grouping	Leader	Role
$63^1/2 = 31$	group	instructor	unity manager
$63^2/2 = 1984$	company	director	justice advisor
$63^3/2 = 125{,}000$	community	chairman	stability counsellor
$63^4/2 = 7.8$ M	culture	judge	progress manager
$63^5/2 = 500$ M	race	king	relationship advisor
$63^6/2 = 31$ B	humanity	messiah	peace counsellor

Table 8.1 Leadership types

The six types of leader govern differently, respecting and answerable to the others. Anyone may volunteer or be nominated to serve as leader in any capacity, seeking election if opposed and maintaining harmonious relationships with their peers. Election candidates issue manifestoes, electorates are grouped by size. This book may be treated as a manifesto for New Age Messiah-ship. I hope to encourage a thoughtful, scientific approach to life's problems and introduce goodness, truth, beauty, peace, love, progress, stability, justice and unity for decision making. Readers are free to pose questions.

8.8 Religion

Using a freely exchangeable international currency, groups, companies or communities may restrict its use by paying wages and salaries in equal amounts of *gold, bronze or silver* money. Gold money buys food and essential clothing, bronze money is taxable and for recreation and silver pays for rent and essential services. Groups, companies and communities decide the allocations. Children's pocket money includes no silver. Bronze money pays penalties and fines. Charities collect gold or silver money.

Communities are the largest decision-making units, judges, kings and messiahs don't assign funds, or plead for peace, love or progress.

They rely on communities, companies and groups to make these decisions. Leaders' straightened personal circumstances mean that they need take employment to keep in touch with reality. Candidates need to keep their own affairs in order whilst prosecuting the leadership function, utilizing their experience to rule by example. Unpaid officials need no offices of state, ceremonies using robes and thrones are staged only for public entertainment. Leaders' private lives are spent in the community.

The New Age citizen's religious creed always distinguishes nine concepts, making the colour of money significant. Holy New Age citizens lead full lives and handle all colours of money. Everyone's personality is a unique combination of the nine concepts, affording them a unique perspective on life. Education teaches morality, structuring life. It isn't a sanctuary from reality.

Although brought up as a Methodist Christian, I profess myself an atheist, Muslim believers have recently protested violently at Christian dominance. Religious faiths arise when a 'guru' achieves prominence through his or her thinking and proposes rules of conduct, *e.g.* Moses' Ten Commandments. Priests strengthen the fabric of society by arranging meetings, youth clubs and church services.

This book explains the origin of life, a mystery addressed by most religions. *Only human* suggested ways to make wise decisions and *What's Natural* proposes food choices to prevent illness. Apart from 'ice It', my ideas aren't new – they integrate available wisdom, both ancient and modern. Monasteries, nunneries and temples provide places to meditate and revise teaching.

8.9 Space

People's rights to space are constrained by the formula 2×6^{2N}. Everyone has the right to $2\ m^2$ for sleeping, to share a field to grow food and to travel unhindered showing a passport at frontiers, Table 8.2.

Area	Description	Role
2 * 63^0 = 2 m^2	bed	individual
2 * 63^2 = 90 m square	plot	group
2 * 63^4 = 30 km^2	community	place
2 * 63^6 = 350 km square	district	culture
2 * 63^8 = 500 M km^2	Earth	race

Table 8.2 Space allocations

Like the congregation sizes introduced in chapter 7, these areas are only guides for assessing human rights; a ‘place’ is the largest negotiable area. Right of access to space shouldn’t depend on racial or cultural divisions. Infants learn the right to roam their local plot. When educated, they achieve freedom to travel throughout their local place, respecting the privacy of citizens’ plots and the secrecy of their beds.

Visiting other places within their culture, they learn of other cultures where commodities may be scarce and with different manners and by-laws. As adults, they earn the liberty to visit other cultures, respecting their customs and laws and keeping the peace. A citizen’s freedom to travel is earned through education. They should respect local traditions abroad and not interfere with others’ cultures. The principles controlling human behaviour are as mysterious as fundamental particles, don’t try to change them.

9 Being Myself

Science texts are inconsequential out of context. This chapter concerns the relevance of my ideas for the late 20th and early 21st centuries. It discusses the differences my new axioms make to consensus science. It's not a declaration of intent to replace existing dogma without debate. I hope its publication will induce the scientific community to assimilate my ideas, not precipitate a revolution. I can't implement my ideas alone but intend promoting them. Few architects are builders. Farmers need help to reap their harvests.

Architects call on teams of men and women with necessary skills, training and experience to cooperate in realizing their designs. Improving the school syllabus is first priority. I seek constraints on leadership, not world government. Any cooperation among men and women takes account of patterns established by their governments.

I conclude by examining the status of science for world citizens and compromising altruism and reverence with enjoying life. Section 4.9 presented my autobiography; autobiographies rewrite history for political purposes. Completing the uncoiling of science could free society from creating political unity through government.

9.1 Pilgrimage

Unless abandoned at birth, one's life is constrained, guided and encouraged through education. I cherish the advice my mother gave, lessons learnt in school and experience with others. They signposted my life's path, afforded a variety of experiences and warned me of pitfalls. Instruction manuals, textbooks and training courses introduced me to other aspects of life. Pilgrimages, journeys such as Chaucer described in *Pilgrim's Progress* afford opportunities to exercise, debate one's beliefs with others and have an adventure to remember later on.

Unlike a package holiday, organization and taking responsibility for the welfare of one's companions teaches love for neighbours. More experience can be gained by extending an open invitation for companions, choosing an unfamiliar route and advertising a worthy cause. Reaching the end is of no consequence, adopting arbitrary gestures such as walking barefoot renders the course more of a challenge and makes accounts of the trip more colourful.

Such a holiday broadens the pilgrims' horizons, leaving them refreshed on returning to everyday routine, with a new set of comparisons to remember. Recounting their experience makes for conversation with old friends, the close encounters with strangers resolves problems. Pilgrimages are educational and fulfilling. Travel books, conversations, religious and social groups may suggest ideas for pilgrimages, adding to life's comedy.

9.2 Intelligence

We can only assess our intelligence and respectability relative to those of others, test scores are reassuring but we learn our potential through experience. Our advantages emerge in competitions and by attaining qualifications and prizes, enhancing our social status. Exercise and study improve intelligence but cost time, effort and money. Society benefits by selecting their brightest progeny and 'hot-housing' them but child prodigies can make greater contributions if given a broad general education. Their insights may challenge the society's precepts and lead to advances.

Tribal initiation rites and university degrees reinforce the status quo, if a gifted individual devises political changes, the public stand to benefit. Democracies are best governed if their educational syllabus is balanced and broad enough to exploit new axioms, inventions and discoveries and keep pace with scientific advances. If those with executive power are qualified, they'll administer justice fairly. State security is put in jeopardy if scientific progress isn't recognized, resulting in corruption. Censorship disrupts the balance between science and society and fosters conflict between arts and sciences.

9.3 Personality

Astrologers contend that personality depends on time of birth, society benefits by fostering diverse personalities through intensive education and guiding career choice. Concentrating on physique or idiosyncrasy detracts from common welfare, life would be dull without comics, athletes and star performers advertising human potential and stirring action in times of peace. Failing to distinguish fact from fiction can distort our judgment. Well endowed individuals may be exploited for good or evil, personalities can either enhance or corrupt the truth.

Privacy, secrecy and modesty, attributes least apparent in film stars, are qualified virtues. Leading open life-styles engenders preferred qualities and sways electorates. Rich, intelligent and physically well-endowed individuals either develop their personalities or lead lives benefitting the community. Exploiting gifted children is outlawed, their curiosity encouraged. Respecting the privacy of people in public life and taxing performance fees help achieve this. Gifted individuals developing altruistic personalities act as missionaries, change society by example, lay new trails for pilgrimage and illuminate new ideas. Followers of personality cults are unfortunate by-products of missionary work.

9.4 Helicore

Helicore (named after Greek Sun-god *helios* or *helix* and the *core* of a computer) was an attempt at modelling human memory, the idea of imitating the action of the brain raises important questions for science and society. Automating everyday life enslaves people to machines, dams harbour malarial mosquitoes and pest control chemicals contaminate food. The governments' role in investigating and controlling the use or abuse of inventions ensures common welfare.

Machine intelligence, drug abuse, toxic chemicals, television, statistical surveys and nuclear power are current areas of conflict between science and society. Inventors, governments and users share responsibility for this conflict. Innovators are well placed to foresee the consequences of their discoveries. Socially responsible scientists plead with governments to respect their views.

The remaining sections of this book document a history of their advice being ignored and my latest research. Leaders have power applying to $63^N/2$ subjects, their scope varies with N. Power is shared between individuals, groups and computers. Giant computers should have limited access to information, using the $63^N/2$ rule as guide. People will become inured to the universal accessibility of personal information about their financial affairs, medical history and politics.

As computers become cheaper, governments can't control them, central information storage gives way to a freely accessible 'cloud' of data. Totalitarian regimes have intimidated their subjects by intercepting their communications and falsifying their records. Doomsday Books provide valuable insights to the lives of our forebears. An informed elite or *Big Brother* state can drive people insane. Artificial intelligence threatens the ignorant and inept, help need be freely available to reassure them.

9.5 Selenium

The history of medicine records many discoveries in fundamental science leading to reduced morbidity, *e.g.* introducing hygienic water supply, immunization and antibiotics. I suggest in section 5.9 that supplementing the human diet with selenium would be a cheap way to prevent heart disease and common forms of cancer. Patients now live longer, happier lives due to organ transplants and intensive care, but at great cost. The cost of maintaining life is measured by the respect accorded to us by others, not everyone can benefit from expensive technologies.

As we live longer, the proportion of pensioners and geriatrics needing community care amongst the electorate increases, promising no gratitude. Our lifestyle is unsustainable unless our social mores changes. We need be less prodigal of resources; physical, intellectual and spiritual activities need encouragement. The long lives of a few healthy, educated intellectuals demonstrate life's potential.

Human lives fall naturally into seven-year periods, passing through the nine concepts: goodness from 0 to 7, truth from 7 to 14, etc, returning to goodness at 63. Many communities celebrate people's 7th, 14th, 21st and 63rd birthdays. The age when childhood innocence ends and adulthood begins has recently declined. Some individuals maintain former standards, take an extended education and reject that change. I propose prolonging innocence to age 31½, half way through the 63-year cycle.

9.6 Cancer

Everyone need restrain themselves to maintain youthful innocence to prevent physical, mental and spiritual cancers. Convenient chemicals can cause cancers; an over-specialized education, mental health issues and excessively young and enthusiastic political and religious leaders contribute to the spiritual condition, secrecy leads to corruption, censorship and scientific understanding are needed, violence and dissent have no place in its creation. There's little advantage in permitting 7-year-olds to smash windows, 14-year-olds to drive automobiles, 21-year-olds to lead armies or 28-year-olds to arbitrate morality.

The economy benefits from granting early majority and sexual freedom but societies developing from such beginnings yield crops of weeds, stifling the growth of those destined to lead it. To ensure the innocence of early life in the New Age, elder citizens must adopt measures to prevent cancerous growths. The respect due to New Age citizens depends on their attainments. Success, attainable by both fool and genius, depends on charm, balance and holiness.

Individuals and societies have made sacrifices through the ages to face life's challenges. Nurturing balance calls for more than using a comprehensive syllabus in schools, education need be tailor-made for every child, matching their abilities with development and deficiencies with training. Their elders will win their pension rights by disciplining and teaching this student army.

A society encouraging foreign travel is cemented in this way, young people can be constrained financially, maintaining a statutory income to age 31½ and denied credit facilities. Exploitation of young people is outlawed. People gain creditworthiness in later life by exercising their talents, having their affairs scrutinized for fraud and avoiding criminal activity, increasing their hope of becoming respected citizens. Freedom and responsibility are granted at age 31½, when individuals would be free to market their talents and exploit men, money and resources. The franchise for government, whatever its democratic basis, would extend from 31½ to 63, nurtured by maintaining control over young and old and balancing teachers and students.

9.7 Leisure

The freedom, license and leisure enjoyed by 31½–63-year old citizens, may be controversial and misunderstood; freedom to form groups enjoins responsibility to do so, not to retire to the isolation of fireside and television. I don't advocate any particular leisure activities. Everyone is free to choose a path for developing their personality as long as it's responsible. Every social malady may be addressed by leisure activities, whether a scientific problem, an area of doubt or art form offering scope for innovation. The leisured classes can extend the educational syllabus by preparing texts and courses and amend scientific laws. Both young and old make suggestions, citizens carry out research and development.

Retirement at 63 is compulsory for laboratory administrators, chairing committees, speeding development and postponing death.

Retired folk expect the advice and cooperation of youngsters and their financial interest in community affairs respected. Age and youth are constrained to a fixed income, discouraging extravagant consumption in old age. During their long and active retirement, citizens set a good example by adopting attractive lifestyles and using their skills to advantage.

Citizens failing to thrive under such freedom may opt to enter a closed order or a cell in the open prison of suburbia, fed a diet of censored entertainment and gambling, rejecting freedom. For them, the 31½-year investment in education failed. Abandoned by society, they may suffer early death through accident or disease. Charitable governments ensure citizens' rights to employment, income and respect, the rest is free to choose.

9.8 Equality

If people were identical, they might be valued and traded with advantage. Inequality isn't inborn, both upbringing and opportunity contribute to it. Enforcing equality of income wouldn't leave everyone equally content. A community is likely to be more successful if led by experts. Legislation under-writes equal rights and controls corruption. The present epoch affords leisure for drafting a constitution based on valid premises to achieve good government. Revolution isn't necessary. Changes shouldn't threaten security, nor challenge other societies.

Every nation can adopt a new system of government. In some systems, voting is a statutory requirement and exploitation of young or old constitutional. Violent overthrow of existing regimes offers no advantage. New schemes may take a generation to stabilize. They're best approached through gradual transitions. Cooperation, self-denial, self- interest and self-help ease their adoption. Within the constraints of their community, readers are free to seek their benefits, they constitute no threat.

9.9 Wisdom

We acquire wisdom throughout life, it's in the nature of knowledge to take time, whether teaching children or writing directions and memoranda. This book's waited over forty years for publication. Its three parts, each with nine chapters, accord with minion logic. Future editions could incorporate further scientific advances to this framework. Each minion coil has a significant meaning.

Meanings of the numbers 1—63, as distinct from 1—9, may eventually emerge. They seem unlikely to elucidate science or advance the frontiers of knowledge. The level of coiling I've unravelled seems sufficiently important to be written into the constitution. Religions proclaim concepts right and wrong. Future theologians may discover rights and wrongs within concepts.

I intend promoting the advantages of my discoveries, issuing new editions in the light of experience. If my book's hard to understand, please contribute ideas to improve it. A glossary explaining technical terms is appended.

Conclusion: Science

Science Uncoiled challenges current scientific dogma. Scientific breakthroughs aren't sacred, scientific progress is gradual, its methods are applicable to everyday life, everything can be improved, dreams may be realized. Don't abandon hope, advance by searching, researching and testing. Be alive, natural and human, give life and you will receive life anew, broadcast your joy in living. E-mail questions to michaeltdeans@gmail.com, I don't promise to answer them. I wish you peace, love and progress, Michael T Deans.

Glossary

Acetylcholine, 5.1 *see* Differentiation
Actin *see* Muscle contraction
Active transport, 4.1, 4.3, 7.6
Adenine, A *see* DNA
Adenosine diphosphate, ADP *see* DNA
Adenosine monophosphate, AMP *see* DNA
Adenosine triphosphate, ATP *see* DNA
Adenyl cyclase *see* Active transport
Adrenalin, 5.2, 5.8
Al Aluminium, Al, 5.6
Ala Alanine, 1.3, 4.4 *see* Amino acids
Albert Einstein *see* Relativity
Alcohol, 4.7, 5.5, 5.7, 6.7
Alois Alzheimer, 4, 5.6
Alzheimer's Disease, 4, 5.6
Amino acids,1.3, 4, 5.5
Andrew Huxley *see* Muscle contraction
Angiotensin *see* Rennin
Antibiotics, 5.3, 5.5, 9.5
Anticodon *see* DNA
Apoptosis *see* cell death
Arg Arginine, 1.3 *see* Amino acids
Aristotle, 7.0
Base pairing *see* DNA
Beads on a string *see* DNA
Beryllium, Be 5.7

Lightning Source UK Ltd.
Milton Keynes UK
UKOW06f2324080316

269804UK00014B/66/P